AF631053

Plantation Crops

NIPA® GENX ELECTRONIC RESOURCES & SOLUTIONS P. LTD.
New Delhi-110 034

Plantation Crops
At a Glance

Authors

Bhani Ram
Mamta Dall
Anil Sharma

NIPA® GENX ELECTRONIC RESOURCES & SOLUTIONS P. LTD.
New Delhi-110 034

NIPA® GENX ELECTRONIC
RESOURCES & SOLUTIONS P. LTD.

101,103, Vikas Surya Plaza, CU Block
L.S.C. Market, Pitam Pura, New Delhi-110 034
Ph : +91 11 27341616, 27341717, 27341718
E-mail: newindiapublishingagency@gmail.com
web: www.nipabooks.com

For customer assistance, please contact
Phone: + 91-11-27 34 17 17
Fax: + 91-11- 27 34 16 16
E-Mail: feedbacks@nipabooks.com

ISBN: 978-81-19254-55-2

NIPA® also publishes books in a variety of electronic formats. Some content that appears in print may not be available in electronic books, and vice versa.

Composed and Designed by NIPA®.

Preface

India is leading producer of plantation crops. Plantation crops are high value commercial crops of greater economic importance and play a vital role in our Indian economy. The main draw back with this sector of crops in India is that major portion of the area is of small holdings, which hinders the adoption of intensive cultivation. Same time plantation crops are good source of foreign exchange.

Contemporary time is time of campaign, competitive examinations have become a regular feature and second thing about its organizations and people are continuously awaring about plantation crops. So as a consequence there are a large number of aspirants who have wished to adorn their career through plantation crops. The present book "Plantation Crops: At a Glance" cover almost all valuable chapters regarding subject matter with Multiple Choice Questions, Matching and Fill in the blanks. The attempt has been made to deliver a lot of subject's information to compete various examination like ARS, NET, JRF, SRF, Ph.D., FCI as well as competition of UPSC and State PCSs for various jobs.

The author considers that the readers will get ample information concerned with the plantation crops. There may, however, be still some shortcoming in this book and authors will be really grateful to receive proposal from readers for incorporation in the next edition of this volume.

Bhani Ram
Mamta Dall
Anil Sharma

Acknowledgements

I dedicate this book to my father and mother, Mr. Mula Ram Dall and Mrs. Guman Devi. I acknowledge with profound gratitude to motivational contribution of Mr. Madan Dall, Lucky Dall, and Ganesh Kumar.

I am also thankful to my M.Sc. research guide Er. P.S. Pandit, who taught me the importance of hardwork and Dr. S.L. Chawala, who always inspired me to dream big.

I am also thankful to Rameshwer Riwaria, Suman Riwaria, Rakesh Goswami, Alka Nirayaniya, Mr Shiv Raj, Mr. Krishan Kumar, Mr. Vijay Kumar, Shyoparkash and Mr. Sanjay Kumar for their contribution in this book.

Bhani Ram

Contents

Chapter 1

Introduction

Botanical Classification of Plantation crops

Common name	Scientific name	Family	Part of uses	Chromosome No.
Plantation Crops				
Areca nut	*Areca catechu*	Palmae / Arecaceae	Nuts	2n=2x=32
Betel vine	*Piper betel*	Piperaceae	Leaf	2n=2x=26
Cashew nut	*Anacardium occidentale*	Anacardiaceae	Kernel / nut	2n=2x=42
Cocoa	*Theobrama cocoa*	Sterculiaceae	Bean	2n=2x=20
Coconut	*Cocos nucfera*	Palmaceae	Endocarp	2n=2x=32
Arabica Coffee	*Coffee Arabica*	Rubiaceae	Beans/ fruits	2n=4x=44
Robusta Coffe	*Coffee canephora*	Rubiaceae	Beans/ fruits	2n=2x=22
Oil palm	*Elacis guineensis*	Palmaceae	Mesocarp	2n=2x=32
Palmyrah Palm	*Borassus flabellifer*	Palmae	Fruit	2n=2x=36
Rubber	*Hevea brasilensis*	Euphorbiaceae	Latex	2n=2x=36
Tea	*Camellia sinensis*	Camelliaceae	Leaves/ shoots	2n=2x=30

The term plantation crops refers to those crops which are cultivated on an extensive scale in contiguous area, owned and managed by an individual or a company. The crops include tea, coffee, rubber, cocoa, coconut, arecanut, oil palm, palmyra and cashew.

- Coconut, cashew nut, cocoa, arecanut, oil palm and palmyra comes under Ministry of Agriculture.
- Tea, coffee and rubber are dealt by Ministry of Commerce.
- Excavations in the Indus valley revealed that spices and herbs have been used even before 1000 BC.
- India is a leading country in plantation crops production.

- Share of India in total world production of tea, coffee, rubber and spices was 25 percent, 3.65 percent, 7 percent and 31 percent in (2012-13) respectively.

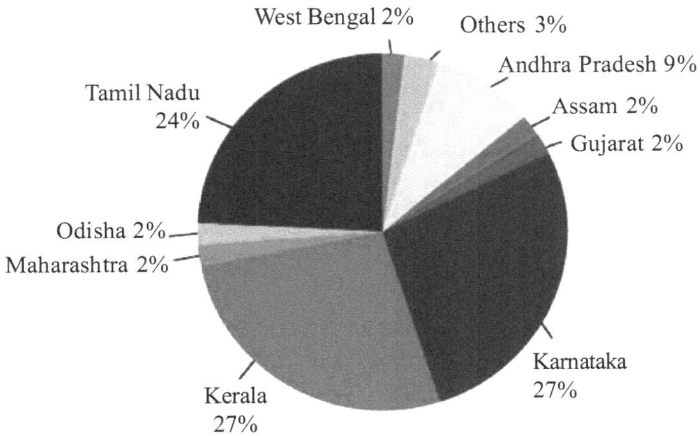

Leading Plantation Crops Producing States

Total area and Production of Plantation Crops in India

Area in '000 ha, Production in '000 MT and Productivity = MT/ha

Area: - 3641000

Production: - 16985

Productivity:- 4.7

Major Plantation Crops Area, Production and Productivity

Area in '000 ha, Production in '000 MT and Productivity = MT/ha

Crop	Area	Production	Productivity
Coconut	2137	15609	7.3
Cashewnut	992	753	0.8
Arecanut	446	609	1.4
Cocoa	66	13	0.2

Production Share of Major Plantation Crops in India (2012-13)

Crop	% Share
Coconut	91.90
Cashew nut	4.44
Arecanut	3.58
Cocoa	0.08

The total production and export of plantation crops in 2013-14

Crops	Production	Export
Tea (Million Kgs)	1209	226
Coffee (Tonnes)	304500	312625
Rubber (Tonnes)	844000	5398
Spices (Tonnes)	5833870	817250

Leading state in plantation crops area, production and productivity

Crop	Leading state in area`	Leading state in production	Leading state in productivity
Areca nut	Karnataka	Karnataka	Karnataka
Cashew nut	Andhra Pradesh	Maharashtra	Maharashtra
Cocoa	Kerala	Kerala	Kerala
Coconut	Kerala	Kerala	Lakshadweep
Coffee	Karnataka	Karnataka	Karnataka
Oil Palm	Andhra Pradesh	Andhra Pradesh	Andhra Pradesh
Rubber	Kerala	Kerala	Kerala
Tea	Assam	Assam	Assam

Crop	Leading country in area	Leading country in production	Leading country in productivity
Arecanut	India	India	India
Cashew nut	India	India	India
Cocoa	Indonesia	Indonesia	Indonesia
Coconut	Indonesia	Indonesia	India
Coffee	Brazil	Brazil	Brazil
Oil Palm	Malaysia	Malaysia	Malaysia
Rubber	Thailand	Brazil	India
Tea	India	India	India

Important Board, Directorates and Organizations

Name of Board, Directorates and Organizations	Establishment Year	Head Quarter	Remarks
Coffee Board of India	1942	Bangalore, Karnataka	Function under the Ministry of Commerce and Industry, Govt. of India.
Tea Board of India	1st April 1954	Kolkata	Function under the Union Govt.
Indian Rubber Board	1947	Kottayam, Kerala	Govt. of India.
Coconut Development Board	1979	Kochi, Kerala	Function under the Ministry of Agriculture, Govt. of India.
Central Plantation Crops Research Institute (CPCRI)	1970	Kasargod, Kerala	Function under the Indian Council of Agricultural Research.
Rubber Research Institute	1955	Kottayam, Kerala	------
Directorate of Cashew Research	1986	Puttur, Karnataka	------
Directorate of Oil Palm Research	1995	Pedavegi, AP	------

Chapter 2

Arecanut

Botanical Name: *Areca catechu L.*

Family : Palmae

Origin : India and South Eastern Countries.

Chromosome no. : 2n=32

Type of fruit : One-seeded berry.

Type of Inflorescence : Spadix

Useable Part : Kernel or endosperm of the seed.

- The betel nut is a nut that chewed as a mild stimulant.
- It is probably used much more frequently than other masticator agents including tobacco, chewing gum and cola.
- The Areca palm is also used as an interior landscaping species.
- Popularly known as Betel nut or Supari, Areca palm, Arecanut palm, Betel palm, Indian nut, Pinang palm, Adike, Betel nut, Fobal, Goorrecanut Palm, Gouvaka, Kamuku, Mak, Paan supari, Tambul, Tuuffel.
- This palm is called the betel tree because its fruit, the arecanut, is often chewed along with the betel leaf.
- The seed contains alkaloids such as arecaidine and arecoline, which, when chewed, are intoxicating and slightly addictive.
- It is grown primarily in Karnataka, Kerala, Assam, Maharashtra and West Bengal.
- Extensively cultivated tropical palm the nuts of which form a popular masticatory in India, the Middle East and the Far East.

- The major arecanut growing countries in the world are India, China, Bangladesh, Indonesia and Myanmar.
- India is the largest producer of arecanut and at the same time largest consumer also.
- Areca nut is an erect, un-branched palm reaching heights of 12-30 m, depending upon the environmental conditions.
- Major states cultivating this crop are Karnataka (40%), Kerala (25%), Assam (20%), Tamil Nadu, Meghalaya and West Bengal.
- Areca nut requires abundant and well distributed rainfall. It grows well within the temperature range of 14-36°C. It can be cultivated up to an altitude of 1000 m.
- The cultivation is mostly confined to 20° North and South of the Equator.
- Laterite, red loam and alluvial soils are most suited. The depth of soil should be at least 1 m with 4.5-7.6 pH.
- Sticky clay, sandy, alluvial, brackish and calcareous soils are not suitable for areca nut cultivation.

Varieties

S.No.	Variety	Growth habit	Shape and size of nut	Chali yield (kg/palm)	Recommended for Local Varieties
1.	Theerthahally Local	Tall	Oblong small	2.60	Malnad areas of Karnataka
2.	Maidan Local	Medium	Round nuts	2.6	
3.	South Kenra Local	Tall	Round, Bold	2.00	Coastal Karnataka & Kerala
4.	Mohitnagar	Tall	Oval to Round, Medium	3.67	West Bengal, Karnataka and Kerala
5.	Srivardhan	Tall.	Oval, Medium	2.20	Raigad and Ratnagiri areas of Maharashtra
Improved Varieties					
6.	Mangala	Semi-tall	Round, Small	3.00	Coastal Karnataka & Kerala
7.	Sumangala	Tall	Oval, Medium	3.20	Karnataka and Kerala
8.	Sreemangala	Tall	Round, Bold	3.28	Karnataka and Kerala

S.No.	Variety	Growth habit	Shape and size of nut	Chali yield (kg/palm)	Recommended for Local Varieties
9.	Sarwamangala (VTL-12)	Semi tall to tall	Oblong, Round, Bold	3.90	Karnataka and Kerala
10.	SAS-1 (Released by UAS, Dharwad)	Tall	Oval Medium	4 .60	Uttarakhand
11.	Vittal Areca	Dwarf	Medium sized oval	2.54	Karnataka and Kerala
12.	Vittal Areca Hybrid – 2 (VTLAH-II):- released by CPCRI Vittal		Medium sized Round	2.64	Coastal belt

- Areca nut is propagated only by seeds.
- Fully ripened nuts having weight of above 35 g should be selected. The weight of seed nuts for red varieties should be 20-25 g.
- The middle bunches should be selected for seed purpose leaving the first and the last bunch. The nuts selected should float vertically with calyx-end pointing upwards when allowed to float on water. These nuts produce the seedlings of greater vigor.
- Planting should be done in May-June in well drained soils.
- Twelve to eighteen month old seedlings are to be selected and transplanted in the main field.
- Selection of seedlings can be based on Selection Index, which is calculated by multiplying leaf number by 40 and sub-starting the seedling height.
- Select seedlings with higher Selection Index Values i.e. if index values range between 50 and 150, select seedlings with higher values.
- Remove seedlings with ball of earth attached to them for transplanting.
- Planting is to be done at a spacing of 2.7x2.7m. While planting with Cocoa, a spacing of 5.4x2.7 m is recommended.
- Pits size of 90x90x90 cm.
- The crop thrives well in humid areas protected well against hot sun and heavy wind. So the site selected should have protection from southern and western sides.
- In low laying areas and South-West monsoon is severe, where water logging is a common planting during September-October after South West monsoon is of the crops.

- The application of NPK fertilizers for adult palms @ 100:40:146 g/palm/year and 12 kg. of each of green manure and compost is recommended.
- In acidic soils, broadcast lime @ 0.5 kg per palm once in two or three years and incorporate into soil during April-May.
- The palms are highly susceptible for sun scorching.
- Sun scorching is mostly seen during October – January. During this period even the stems of young palms have to be protected.
- Planting of banana along with areca nut is a very good practice where banana provides shade to young areca nut plants at the same time giving additional revenue to the farmer in the form of bunches.
- Mimosa in visa, *Stylosanthes gracilis. Calapogonium muconoides* and *Pueraria javanica* are suitable as cover crops.
- The seed rate required per hectare for Mimosa, Stylosanthes, Calapogonium and Pueraria is 15 kg, 9 kg, 11 kg and 11 kg, respectively.
- Banana, pepper and cocoa can be grown in inter-spaces as mixed crop in coastal Karnataka and Kerala.
- Acid lime and betel vine are suggested in West Bengal and Maidan parts of Karnataka.
- In a pure crop of arecanut only 31% of the land area is utilized by the main crop and remaining 69% area is available for growing of inter and mixed crops.
- Harvested fully ripe nuts alone are harvested for preparation of *Chali.*
- The harvested nuts can be sun dried for about 45 days. Proper drying of the nuts is important to prevent fungal infection of the nuts in the drying yard.
- Areca nut is harvested in 3-4 harvesting period in a harvesting season at an interval of 30 days.

Harvesting: The palm start bearing from 4-5 years.

- The plant is. monoecious, producing both male and female flowers on the same tree.
- The spadix of a grownup palm produces on an average 294 female flowers.
- The colour of the fruit during its growth changes from green to different shades of yellow and red during ripening.

- Nut matures after 8-9 months after fertilization, 4-5 spadix are produced by the palm per year.

Tender nut Arecanut Processing

1. **Dehusking of raw nuts to get the Kernel:-** One of the most important activities in the entire arecanut processing is this step. The raw fruit has to be peeled in order to get the kernel. This has to be done within a day or two after harvesting. Otherwise, the cutting will not be easy at all.
2. **Boiling the kernel:-** Immediately after dehusking, the raw nuts have to be boiled with high temperature for a minimum of 3-4 hrs. This process gives attractive reddish colour to the processed nuts.
3. **Sun drying the boiled kernel:-** After boiling the arecanut kernel has to be sun dried for about 4 to 7 days. This is the last step of arecanut processing. Once the drying process is over, it can be packed in and kept for any number of days before selling it in the retail market.

Processing of ripe nuts (Chali production):- For production of chali the ripened nuts are harvested from the tree. After harvesting the fruits are separated from the bunches and they are sun dried for 40 to 45 days on a clean cement/ tiled floor are a short pendal erected near the threshing yard for drying of ripe nuts. During drying process the fruits are turned over periodically to facilitate uniform drying.

Important Insect Pests of Arecanut

1. Scales (*Coccus hesperidum* and *Coccus acutissimum* and *Saissetia* sp.)
2. Mealy bugs (*Icerya seychellarum and Pseudococcus cryptus)*
3. Arecanut nut borer *(Tirathaba nrrufivena)*
4. Snails (*Achatina fulica)*
5. Spindle bug (*Carvalhoia arecae*)
6. Mites (*Raoiella indica, Oligonychus indicus*)
7. Root grub (*Leucopholis lepidophora, Leucopholis coneophora* and *Leucopholis burmeisteri*)
8. Pentatomid bug (*Hapylomorpha marmoreal*)
9. Inflorescence caterpillar (*Tirathaba mundella* and *Batachedra* sp.)

Important Disease

1. Bacterial Leaf Stripe: This is caused by *Xanthomonas campestris* pv *arecae.*
2. Bud rot (Crown rot): *Phytophthora arecae.*
3. Koleroga (Mahali): *Phytophthora areceae / Phytophthora meadi.*
4. Anabe roga / Foot Rot: *Ganoderma lucidu.*
5. Leaf spot: *Pestalotia, Phylosticta* and *Colletotrichum gloeosporiodes.*
6. Yellow leaf disease: This is caused by *Phytoplasma* and transmitted by the vector, *Proutista moesta.*
7. Inflorescence dieback and Button shedding

Physiological Disorder

1. **Nut splitting**: The disease characterized by the cracking of fruits known as "*Anduadakke Roga*" in Kannada. It is due to Boron deficiency.
2. **Sun Scorch (Stem breaking)**: It is due to high sun light.
3. **Band disease:** It is due to Copper deficiency.

Chapter 3

Cashew Nut

Botanical Name : *Anacardium occidentale* L.

Family : Anacardiaceae

Origin : North East Brazil in Latin America.

Chromosome No. : 2n=2x=42

Type of fruit : Droup.

Type of Inflorescence : Panicles (Both male and hermaphrodite flowers occur on the same inflorescence)

Useable Part:- Kernel.

- The genus *Anacardium* is a native to Latin America and has a primary centre of diversity in Amazonia and secondary one in the Planalto of Brazil.
- Cashew is an evergreen tree of tropics; this is cultivated in more than 28 countries in tropical region on either side of the equator.
- Cashew is used in Goa for preparation a popular alcoholic beverage called "Feni".
- CNSL (Cashew nut shell liquid) is used in paints, textile, resin industries and ship building industries and it is considered as an important by product of high trade value.
- Cashew was introduce in India on the coasts of Goa in 1570 AD by the Portuguese for soil conservation, cashew has adapted very well to Indian conditions.
- Cashew is considered as "A waste land crop".

- India is one of the major producers of cashew nut in the world; it imports huge quantity of raw nuts.
- Brazil, Vietnam, Tanzania, Mozambique, Indonesia, Sri Lanka, Australia and Nigeria are the other countries making significant contribution to the world's cashew production.
- In India, Kerala, Karnataka, Goa and Maharashtra on West Coast Tamil Nadu, Andhra Pradesh, Orissa and West Bengal on East coast and Tripura, Meghalaya, Assam and Manipur in the North Eastern Hill Region are the major cashew growing states.
- Cashew is a small-medium-sized tree of Savannah habitat that can grow from 6 - 15 m tall with canopy diameter ranging from 10 - 15 m.
- Flowering normally occurs following the growth flush at the end of the wet season, but its timing and duration are strongly influenced by temperature.
- Flowers are produced at the end of the new shoots. Thus flowers and fruit are borne on the outer extremity of the canopy.
- Both male and bisexual flowers are produced, with male flowers predominating for the first month.
- Pollination is mostly by insects.
- After pollination, the fruit takes 6 to 8 weeks to develop. The nut develops first and the apple fills out during the last fortnight before nut drop occurs. Nut drop continues for 6 to 8 weeks.
- Maharashtra leads in production with 2.25 lakh metric tones and productivity of 1500kg/ha., while Andhra Pradesh leads in area (1.82 lakh hectares) with production and productivity of 1.12 lakh MT and 920 kg/ha respectively.
- The cashew is a hardy drought-resistant tropical or subtropical tree.
- Cashew is cultivated in varied climates but most suitable hot humid tropical conditions and between the latitudes of around 25° North or South of the equator. Although cashews can withstand high temperatures, a monthly average temperature of 27°C is considered optimum but in summer with day temperature of 30-38 °C and relative humidity of more than 50 per cent are congenial for its cultivation. It comes up well upto an altitude of 700m from sea level. Young trees in particular are very susceptible to frost, and cool spring conditions tend to delay flowering.

- Cashew can be cultivated in almost all types of soils ranging from coastal sandy and red lateritic soils to red sandy loam soils with 5.8 to 6.8 pH range and soils with fertility status.

Varieties

Area	Varieties
Maharastra	Vengurla-1,Vengurla-2,Vengurla-3,Vengurla-4,Vengurla-5,Vengurla-6,Vengurla-7, Vengurla-8.
Kerala	Akshaya (H-7-6), Amrutha (H-1597), Anagha (H-8-1), Anakkayam-1 (BLA-139-1), Dhana (H-1608), Dharasree (H-3-17), K-22-1, Kanaka (H-1598), Madakkathara -1 (BLA-39-4), Madakkathara-2(NDR-2-1), Priyanka (H-1591), Sulabha (K-10-2), Raghava, Damodara, Poornima.
Andhra Pradesh	BPP-1, BPP-2, BPP-3, BPP-4, BPP-5, BPP-6, BPP-8(H2/16).
Karnataka	Chintamani-1,NRCC-1,NRCC-2,Ullal-1,Ullal-2,Ullal-3,Ullal-4,UN-50.
Tamil Nadu	Vridhachalam-1, Vridhachalam-2, Vridhachalam-3, VRI 4, VRI (CW) H1.
Goa	Goa-1, Vengurla-4, Vengurla7, Vengurla-8, Bhaskara (Goa 11/6).
Orissa	Bhubaneswar-1, BPP-8, Dhana.
West Bengal	Jhargram-1, BPP-8.
North Eastern Hill Region	Ullal-3, Ullal-4, Vengurla-1 and Vengurla-4.

Varieties and their characteristics

Variety	Characteristics
Vengurla-4	It is an early bearing hybrid variety of the cross between Midnapur Red X Vetore-56. Average nut yield is about 15kg per tree. It yields medium nuts(> 7.5g) and red apples are of medium size.
Vengurla-7	It is a hybrid variety of the cross Vengurla-3 X M 10/4. Nuts are bold and yellow coloured apples are medium in size.
Vengurla-8	It is hybrid variety of the cross Vengurla-4 X M10/4. Nuts are medium bold and apples are yellow and medium sized.
Bhaskara (Goa 11/6)	It is a colonel selection from a mother tree for higher nut yield (10-12 kg/tree) and tolerance to tea mosquito bug, located in Gaondongrim village, Cancona taluka of Goa. Nuts are of medium size (7kg) with 30 per cent of shelling.
Madakkathara-I (BLA-3 9-4)	Cashew apple colour is yellow, shelling percentage is 26.8, nut yield per tree is 13.8kg and recommended regions for cultivation is Kerala, West Bengal, Manipur and Andaman & Nicobar Islands.
Madakkathara-2 (NDR-2-1)	Cashew apple colour is red, shelling percentage is 28, nut yield per tree is 17kg and recommended regions for cultivation is Kerala.
K-22-1	Cashew apple colour is red, shelling percentage is 23, nut yield per tree is 13.2kg and recommended regions for cultivation is kerala.

variety	Characteristics
Kanaka (H-IS98)	The nut size of the variety is found to be small, cashew apple colour is yellow, shelling percentage is 30.58, nut yield per tree is 12.8kg and recommended regions for cultivation is Kerala.
Dhana (H-1608)	Cashew apple colour is yellow, shelling percentage is 27.08, nut yield per tree is 10.66kg and recommended regions for cultivation is Kerala.
Priyanka (H-1591)	Cashew apple colour is yellowish red, shelling percentage is 26.54, nut yield per tree is 16.90kg and recommended regions for cultivation is Kerala.
Vridhachalam-I (M.IO/4)	Cashew apple colour is yellow, shelling percentage is 28.5, nut yield per tree is 7.12kg and recommended regions for cultivation is Tamil Nadu.
Vridhachalam-2 (M.44/3)	Cashew apple colour is pinkish yellow, shelling percentage is 28, nut yield per tree is 6kg.
Vridhachalam-3 (M.26/2)	Cashew apple colour is pink, shelling percentage is 29.1, nut yield per tree is 14.19kg and recommended regions for cultivation is Tamil Nadu.
H-1597	Cashew apple colour is yellow, shelling percentage is 31.58, nut yield per tree is 17.6kg and recommended regions for cultivation is Kerala.
H-1600	Cashew apple colour is red, shelling percentage is 27.27, nut yield per tree is 13.48kg and recommended regions for cultivation is Kerala.
H-1610	Cashew apple colour is yellow, shelling percentage is 26.09, nut yield per tree is 10.62kg and recommended regions for cultivation is Kerala.
Goa-l(Balli-2):	It is a colonel selection from a high yielding plus tree located in Balli village of Quepem Zone of Goa. It is a mid season variety bearing medium nuts (7.6-7.8g)). Shelling percentage is 29-31. Yellow apples are juicy and bigger in size (> 70g). This variety yields 9-12 kg of raw nuts per tree.

- Soft wood grafting is the commercial method for propagation. Most suitable time for grafting is June to October.
- Air-layering is easiest methods for its propagation. However, inadequate root system with low number of initial roots, poor field establishment, uprooting of air layers during the cyclone and susceptibility to drought conditions makes it unsuitable for commercial multiplication.
- The scion shoot should be healthy, non-flowering matured shoots of 4-5 months of current growth.
- Root stock seedlings should be 7- 8 weeks old.
- Top working is used in Cashew tree for rejuvenating /converting the unproductive cashew trees into trees of high yielding varieties.

- Planting distance is 7.5m x 7.5m.
- Pit size:- 60cm x 60cm x 60cm.
- Terracing and Bunding is important operation in cashew field.
- Pruning time:- Pruning of cashew plants should be done during August–September.

Manures and fertilizers

Age	Compost/FYM (KG)	Nutrients (g/tree/year)		
	N	P_2	O_5	K_20
1.	15	250	50	50
2.	30	500	100	100
3.	40	750	200	200
4 year & onwards	50	1000	300	400

- Cashew plants start bearing 3 years after planting.
- Cashew nuts are harvested during February–May.
- Only fully mature nuts should be harvested.
- The nuts are picked after they fall off from the trees.

Yield : Yield is about 2kg nuts at the age of 3–5 years, 4kg (6–10 years) to 5–10kg at 11–15 years and more than 10kg at 16–20 years.

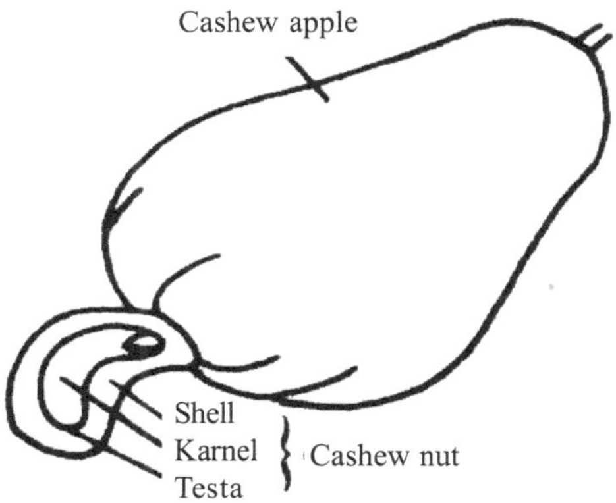

Important insect pests of Cashewnut

S.No.	Insect	Scientific Name	Character
Major insect-pests			
1.	Tea mosquito bug (TMB)		Incidence of Tea mosquito bug on cashew occurs at new flushing, flowering and fruiting stages during September to April and causes crop looses ranging from 20-60 per cent. The damage is caused by adult bugs as well as by immature stages (nymphs) which suck the plant sap from the tender flushes and succulent shoots, flower panicles, growing nuts and apples.
2.	Cashew stem and root borer (CSRB)		This is the most dreaded pest of cashew trees, the immature stages (grubs) of which feed on the bark of trunk and roots and out rightly kill the tree within few days.
Minor pests			
3.	Leaf and blossom	*Lamida moncusalis* L.	This pest attacks the young leaves of new flush webber and inflorescences. The caterpillars form the web of young leaves and flower panicles and feed on these from inside as a result of which, the webbed shoots totally dry up.
4.	Leaf miner	*Acrocercops syngramma* M.	Larvae of this insect mine through the young leaves of new flushes. In the mined laminal areas the epidermal layer gets separated from the leaf tissues in patches giving white blistered patches.
5.	Apple and nut	*Thylocoptila* sp.	The larvae of this pest bore the region of juncture borer where the nut is attached to apple and start feeding on tissues The incidence can be conspicuously seen by the presenceof frass and excreta of the larvae at the tunneled place.
6.	Mealy bug	*Ferrisia virgata*	

Important disease of cashew nut

Disease	Causal organism	Symptoms	Control measure
Anthracnose	*Colletotrichum gloeosporoides* (Fungus)	Water-soaked lesions on leaves, twigs, flowers or young apples which develop into orange-brown or red lesions	A protective coating of copper-based fungicide on susceptible parts of plant can prevent the disease; fungicide should be applied when buds begin to expand through to fruit set but are not required during dry periods
Die back or Pink disease		Prune the affected shoots just below the affected portion	Bordeaux paste. Spray 1 % Bordeaux mixture or any copper fungicide like Blitox or Fytolan 0.25 % twice i.e. in May - June and again in October as a prophylactic measure.
Angular leaf spot	*Septoria anacardii* (Fungus)	Angular cream colored lesions with dark-brown margins on leaves of seedlings; angular black lesions with chlorotic halos on mature trees; defoliated seedlings	
Black mould	*Pilgeriella anacardii* (Fungus)	Chlorotic spots on upper surface of leaves which spread to lower surface as infection progresses; dark-brown to black fungal patches on leaves; leaves shrivelling and dropping from plant. Damage most severe on dwarf cashew varieties	

Cashew nut processing

Cleaning

All raw nuts carry foreign matter, consisting of sand, stones, dried apple etc. The presence of foreign matter in the roasting operation can be avoided by cleaning the nuts. The raw nuts can be sieved by hand using a ¾ inch mesh sieve.

Soaking

The next stage is to soak the nuts in water to avoid scorching them during the roasting operation. After being left to stand for about ten minutes, the water should be drained off. The nuts should then be left for periods of not less than four hours in order to allow the water left on the surface of the nuts to be absorbed. The process of covering the nuts with water, draining and standing should be repeated with the same nuts about three times until a moisture content of 9 per cent is reached.

Roasting

The application of heat to the nut releases the nut shell liquid and makes the shell brittle which facilitates the extraction of the kernel when breaking the shell open. Three methods of roasting exist:

Open pan: An open, mild steel, circular dished pan of around 2 feet in diameter is supported on a basic earth fireplace. When heated, 2-3lbs (1kg) of nuts are placed on the pan at one time and stirred constantly. The CNSL starts to exude and then ignites. This produces a long flame and black smoke. After approximately two minutes, the pan is dowsed and the charred, swollen and brittle nuts are thrown out of the pan. The moisture evaporates quickly leaving the nuts ready for shelling.

Drum roasting: The idea of continually feeding the nuts into a rotating drum over a fire developed from the pan method. A slight horizontal slope in the mounting ensures the movement of the nuts through the drum. The drum is pierced so that the flames touch the nuts and the smoke is controlled by a hood and chimney arrangement. The nuts are dowsed using a continuous spray. This process was further modified by using the heat from the burning CNSL eing harnessed to roast the nuts some more. The roaster consists of a contained helical screw which moves the burning nuts at a controlled rate. The design was a distinct improvement, with little fuel being consumed and there being greater control on the roasting time.

'Hot oil' method: The principle employed in this method is that oil-bearing substances i.e. the shells, when immersed in the same oil at high temperature, will lose their oil, thus increasing the volume of the oil in the tank. For this method, conditioning becomes important. The equipment consists of a tank of CNSL heated to a temperature of 185-190°C by a furnace underneath and a wire basket used to hold the nuts for immersion into the tank. The depth of the basket must be sufficient so that the rim remains well above the oil during the roasting. Immersion time can range from 1½ to 4 minutes. About 50% of the liquid is extracted from the nuts. Draining trays are needed at the end of the tank for the roasted nuts to dry and the residue oil can be returned to the tank. Caution must be taken not to heat the tank to over 200°C because at this point polymerization of the CNSL takes place. The temperature can be maintained by continuous firing. The tank should be emptied and cleaned after each day's roasting. The life of a tank made of an eighth inch thick mild steel plate should exceed one and a half years and can be constructed locally with welding facilities.

Shelling

The objective of shelling is to produce clean, whole kernels free of cracks. In India, this operation has always been done manually. Generally two type of shelling: 1. Manual, 2. Mechanical.

Separation

After shelling, shell pieces and kernels are separated and the unshelled nuts are returned to the shelling operation. Usually blowers and shakers are used to separate the lighter shell pieces from the kernels. The greatest problem is to recover small pieces of kernel sticking to the shell. This is usually done manually from a conveyor belt used to carry all the sorted semi shelled nuts.

Pre-grading

Pre-grading can be done before or after drying the kernels and may greatly reduce the final grading work. For large-scale processors pre-grading can be done mechanically, separating mainly the whole from the broken kernels and sometimes separating the different size groups of whole kernels.

Drying

The shelled kernel is covered with the testa and to facilitate removal, i.e. to peel in order to produce the blanched kernel, the shelled kernel is dried. This also protects the kernel from pest and fungus attack at this vulnerable stage. The moisture content is approximately 6% before drying and 3% after.

Drying method

1. Open Sun drying
2. Furnace drying
3. High-volume Furnace drying

Peeling

At this stage, the testa is loosely attached to the kernel, although a small amount of kernels may have already lost the testa during the previous operations. Manual peeling is done by gentle rubbing with the fingers. Those parts still attached to the kernel are removed by the use of a bamboo knife. One person can peel about 10-12kg of kernels per day.

Grading

The grading operation is important as it is the last opportunity for quality control on the kernels.

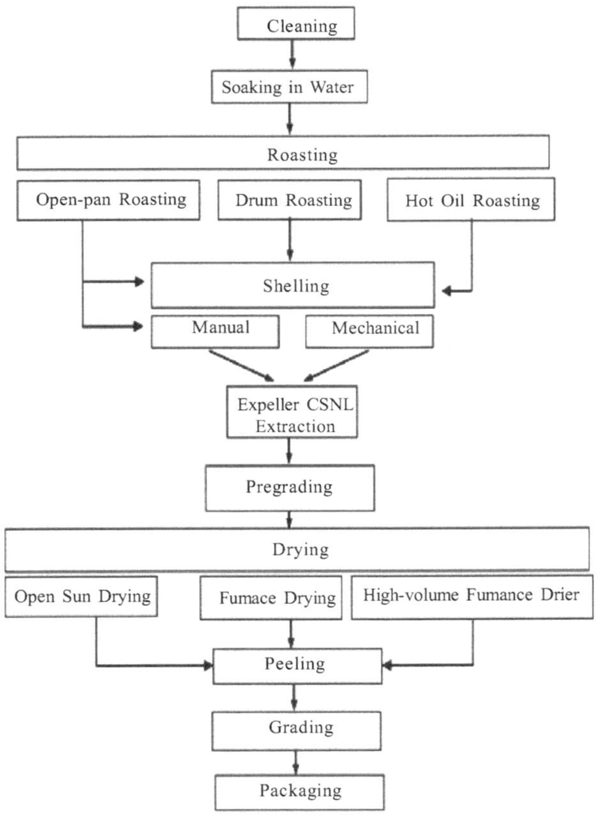

Chapter 4

Cocoa

Botanical Name: *Theobroma cacao* L.

Family : Sterculiaceae

Origin : Central and South America.

Chromosome No : 2n=20

Type of fruit : Drupe.

Type of Inflorescence : Dichasial (cyme)

Useable Part : Seed

- Cocoa also known as "Food of the gods".
- The main cocoa producing countries are Ghana, Nigeria, Ivory Coast, Brazil , Malaysia and Cameroon.
- The cultivation of cocoa on a large scale started in India in the early 1970s.
- It is a small [4–8 m (13–26 ft) tall] evergreen tree.
- Its seeds, cocoa beans, are used to make cocoa mass, cocoa powder, and chocolate.
- It is mainly grown as a mixed crop in coconut and arecanut plantations and as under-storey crop in partially cleared forests.
- The flowers are produced in clusters directly on the trunk and older branches; this is known as cauliflory.
- 159 Exotic cultivars are maintained in CPCRI, Kasaragod, Kerala
- Cocoa flowers are Self-incompatible.

- Less than 10 per cent of cocoa flowers are successfully pollinated, and only about 2 per cent of those (*i.e.* 1 in every 500 flowers) grow into a mature fruit.
- While pollination is normally carried out by insects, higher rates of pollination can be achieved by hand - but cherelle wilt prevents much higher rates of fruit set from being obtained.
- Flowers open in the early morning and hand pollination accomplished by using a small artists paint brush that is first placed in contact with the anthers of an open flower and then placed in contact with the stiga of another flower.
- Cherelle wilt is a natural fruit thinning mechanism whereby the young fruit, called cherelles, stop growing, turn black and shrivel but do not fall off the tree. Cherelle wilt is not a disease.
- Boron has been found to increase pod set and reduce cherelle wilt. Hence, in some places, boron is applied to trees as an annual foliar spray.
- The pod contains 20 to 60 seeds, usually called "beans", embedded in a white pulp.
- Nutritional composition per 100g cocoa powder: Carbohydrate 16.5g, Protein 21.5g, Fat 11g, Dietary fiber 34g, Polyphenols 7-18g, Theobromine 2.5g and Caffeine 0.1g.
- Kerala accounts for 71 per cent of the area and 80 per cent of the production.
- Andhra Pradesh has the second highest area, Karnataka ranks second in production.
- Cocoa is a crop of the humid tropics.
- Cocoa is grown from sea level up to an elevation of about 500m. It comes up best upto 300m above mean sea level.
- Ideally, cocoa requires a minimum rainfall per annum 1500-2000 mm.
- It grows within a temperature range of 15-39°C and optimum temperature is around 25°C. It cannot be grown in areas where the minimum temperature falls below 10°C and the annual average temperature is less than 21°C.
- Cocoa requires deep and well-drained clay loam and sandy loam soils. It thrives well on wide range of soil types with pH ranging from 4.5-8.0 with optimum being 6.5-7.0.

- Commercial cocoa has two major types, 'Criollo' and 'Forastero' which differ mainly in the colour of the cotyledons. That are under:

    1. ***Criollo***: its beans are plumpy and white when fresh and turns cinnamon coloured on fermentation. The dark red 'Criollo' pods have a rough surface, pronounced point and a thin wall. It is possesses a bland flavour with bitter taste. 'Criollo' ferments fully in three days. The 'Criollo' variety has a lower adaptability and less yield potential. It is highly susceptible to diseases. Hence, its cultivation is not encouraged anywhere in the world.

    2. ***Forastero***: This variety forms most of the commercial cocoa of the world. 'Forastero' beans are flat and the colour changes from purple when fresh to dark chocolate brown on fermentation. 'Forastero' pods are green when immature and yellow when ripe, thick walled, melon-shaped with rounded ends and smooth inconspicuous ridges. It is require six days for ferments.

    3. ***Trinitario***: It is hybrid between 'Criollo' and 'Forastero' with highly variable pod characteristics.

    4. ***Amelonado***: It is a 'Forastero' type with a melon-shaped pod, cultivated in West Africa.

    5. ***Amazon***: It is a relatively new type collected from the Amazon forests which has good vigor and high yield potential.

Varieties

S. No.	Variety	Character
1.	CCRP 1	Pods medium sized, green which changes to yellow on ripening. Mature pods weigh 385 g, with 46 beans and 0.8g oven dry bean weight. Average yield 56 pods / tree / year with a potential yield of 72 pods.
2.	CCRP 2	Pods large, green in colour changing to yellow on ripening. Mature pods weigh 311.3g with 45.5 beans and 0.9g oven dry bean weight. On an average a tree yields 54 pods/plant /year.
3.	CCRP 3	Pods medium sized, large in number, green in colour changing to yellow on ripening. Mature pods weigh 240.6g with 42.3 beans and 1.0g oven dry bean weight. On an average a tree yields 68 pods/plant /year.
4.	CCRP 4	Pods large, purple tinged turning yellow on ripening. Mature pods weigh 402 g with 45beans and 1.1 g oven dry bean weight. Average yield is 66 pods 1 tree/year with a potential yield of 93 pods.
5.	CCRP 5	Pods large, elliptical, green when immature turns yellow on ripening. Mature pods weigh 425 g with 42 beans and 0.8 g oven dry bean weight. Average yield is 38 pods/tree/year with a potential yield of 55 pods.

S. No.	Variety	Character
6.	CCRP 6	Pods are very big, green turning to yellow on ripening. Maturepods weigh 895 g with 48 beans and 1.9 g oven dry bean weight. Average yield is 50 pods/ tree/year with a potential yield of 180 pods.
7.	CCRP 7	Pods large, elongated, green turning to yellow on ripening. Mature pods weigh 526 g with 47 beans and 0.9 g oven dry bean weight. Average yield is 78 pods / tree with a potential yield of 95 pods.
8.	CCRP 8	Hybrid between CCRP 1 x CCRP 7. Pods green, medium sized, turning yellow on ripening. Mature pods weigh 389 g with 49 beans and 0.88 g oven dry bean weight. Average yield 90 pods/tree giving 11.40 kg wet beans.
9.	CCRP 9	Hybrid between CCRP 1 x CCRP 4. Trees are self incompatible. Pods green, medium sized, turning yellow on ripening. Mature pods weigh 370 g with 37 beans and 0.8 g oven dry bean weight. Average yield 106 pods/tree giving 8.97 kg wet beans.
10.	CCRP 10	Hybrid between CCRP3 x GVI 68. Pods green, medium sized turning yellow on ripening. Mature pods weigh 332 g with 41 beans and 1.1 g oven dry bean weight. Average yield is 80 pods/ tree giving 8.15kg wet beans.

- Cocoa Propagated by seed, softwood grafting and cleft grafting.
- Cocoa is a self- incompatible crop.
- Cocoa can be grown in areca nut and coconut gardens as a mixed crop. It can also be planted in forest lands by thinning and regulating the shade suitably.
- It can be planted either in the beginning of the South-West monsoon (May-June) in low rainfall areas or at the end of the monsoon (September) in high rainfall areas.
- Cocoa seedlings are to be planted in pits of 60 cm^3 filled with compost.
- Cocoa is planted in areca garden planted with a spacing of 2.7m x 2.7m. While establishing new garden a spacing of 3.3x3.3m can be followed for both areca and cocoa.
- Cocoa is sensitive to water stress.
- A dose of 100:40:140g of N, P_2O_5 and K_2O/plant/year should be given.
- Cocoa plants are grown under the shade of arecanut and coconut plantations. It is therefore necessary to regulate the canopy size and shape of plants so that the main crop is not affected. Proper and systematic pruning is essential in cocoa cultivation.
- Pruning in mature cocoa includes two types:- Sanitary pruning and Structural pruning.

- Pruning is usually done annually in July-August.
- Cocoa produces flowers from the second year of planting onwards and the pods take about140-160 days to ripen.
- Each pod will have 25-45 beans embedded in a white pulp (mucilage).
- Generally two main crops in a year i.e., September-January and April-June though off-season crops may be seen throughout the year especially under irrigated condition.

Cocoa Processing

Fermentation

Fermentation of cocoa beans is essential to remove the adhering mucilaginous pulp to develop flavour and aroma precursors, reduce bitterness and kill the germ of the seed and to loosen the tests. Different methods of fermentation normally followed are

1. Box method
2. Heap methods
3. Tray methods
4. Basket methods.

Box method: This method is more applicable in large estates or central fermentations. The boxes of 60 cm x 60 cm x 45 cm made of wood and having reapers at the bottom to allow the sweating from the pulp to drain out and provide aeration are used. The boxes could be arranged in tiers for transferring beans from one to the next in line below. Two detachable wooden planks are provided on one side of the box for transferring (mixing) the beans by removing the planks. The beans are loaded on fermentation box and covered with banana leaves or gunny bags. The mixing of beans is effected while transferring to the next box after 24 hours. The mixing is done to facilitate uniform fermentation and to maintain proper temperature, moisture and aeration during fermentation. The temperature of the fermentation mass will rise to 42-48°C after about 48 hours of fermentation. Again transferring of beans is done at 72 hours of fermentation and the final transferring is done after another day's fermentation. A total of six days (144 hours) are required to complete the fermentation.

Basket method: Bamboo or cane baskets of suitable size could be used for fermenting small quantity of beans. One or two layers of banana leaves are placed at the bottom with provision to drain the sweating. The basket is filled with the beans and the surface is covered with banana leaves. A small weight is placed over the banana leaves. The basket is placed over a raised surface to facilitate drainage of the sweating for one day. Later the basket is covered with thick gunny bags. The beans are mixed thoroughly on the third and fifth days and again covered with gunny bags.

The fermentation will be completed at the end of the sixth day and the beans are withdrawn for drying.

Drying

After fermentation, the beans can be dried by sun-drying or artificial drying using electric oven. The fermented cocoa beans have considerable moisture (55-69%) and the drying rate is dependent upon temperature and the airflow.

Sun drying: Sun drying should be adopted as far as possible, as it gives superior quality produce when compared to that by artificial drying. The fermented beans are spread in thin layer over a bamboo mat or cement floor and dried for 5-6 days. The beans are to be stirred from time to time for uniform drying. The moisture content of well-dried beans is around 6 to 7 percent.

Artificial drying (Electric oven) : During the monsoon period, artificial drying is done using Electric oven. The duration of artificial drying varies from 48-96 hours. The drying of beans at high temperature should be avoided as it results in low quality end product. Slow drying in the initial stage gives better quality beans. Mould growth is to be prevented during drying as it affects the quality of the beans. In the electric oven beans are to be dried for 8-10 hours at 50-55 ^{0}C for the first two days followed by continuous drying at 60^{0}C. The total drying period will be 72-96 hours. The beans are to be stirred at regular intervals for uniform drying and to prevent clump formation.

Dry beans grading and storage: The dried beans after cooling to room temperature should be cleaned before storage.

Processing product and process

Bean blending:- On the basis of the analysis of the individual bean lots, an optimal blend is prepared. In this way, fluctuating characteristics can be reduced or evened out before the beans are further processed. An alternate approach is to process specific lots of beans and blend the resulting cocoa liquors.

Cleaning, breaking and winnowing:- The actual production process starts with the following three steps: cleaning, breaking, and winnowing. Their objective is to obtain clean, broken, de-shelled kernels (nibs). The beans are sieved, foreign matter is removed and the clean beans are then broken to loosen the shells from the nibs. After the breaking step, the product is sieved into a number of fractions to reach optimal separation during winnowing. These fractions then go to the winnowing cabinets where the "lighter" broken shell is removed by a stream of air. The breaking and winnowing steps separate the essential ingredient of the cocoa bean, the kernel, most often described as the nib, from its shell. Strong magnets remove magnetic foreign matter from the nib. The nib may then be stored, awaiting further processing. The separated shell is often sold to agricultural mulch or fertilizer producers.

Sterilization and alkalization:-The microbiologically contaminated nib is sterilized in a batch or a continuous process by wetting and heating with steam. After sterilization, the nib can be roasted directly (natural process) or can be alkalized first (Dutching process). Alkalizing or Dutching consists of treating the cocoa nibs with an alkali solution such as potassium or sodium carbonate. It is practiced primarily to modify the flavor and color of cocoa powder or cocoa liquor.

Roasting:-The roasting process has the objectives of reducing the water content and further developing flavor. Roasting is particularly important to the final flavor because the nib's flavor is formed from the precursors that developed during fermentation.

Nib grinding:-The roasted nib is typically ground in a multi-stage process. During grinding, the broken kernels change from a solid to a fluid mass of cocoa particles suspended in cocoa butter. This is due to the high fat content of the bean: About half of the nib is fat. Grinding breaks up the cell structure of the cocoa nibs and releases the cocoa butter.

Cocoa liquor :- After the last stage of the grinding process, the mass is passed through sieves and over strong magnets to remove any remaining coarse cocoa or metal particles. This finely ground fluid mass, the cocoa liquor, can either be stored in tanks to await pressing or shipped and used by chocolate manufacturers for further processing into chocolate.

Pressing:- Cocoa butter constitutes about half the weight of the cocoa nib. This fat is partially removed from the cocoa liquor by means of hydraulic presses applying pressures as high as 450 kg/cm2. Depending upon the pressing time and the setting of the press, the resulting cakes may have a fat content of 10 to 24 percent.

Cocoa cake:- After pressing, the cakes are broken into kibbled cake. The pressing operation is microbiologically vulnerable, as this is the only part of the process when the product is not in a closed system and is thus exposed to the surroundings. Hygienic procedures are therefore of particular importance in the pressing department. Kibbled cake is typically stored by fat content and degree of alkalization and may be blended before pulverization to obtain the desired type of cocoa powder. The cocoa butter is filtered and stored in tanks.

Cocoa powder:-The powder grinding lines pulverize cocoa cake particles into the defined fineness levels. After pulverization, the powder is cooled so that the fat of the cocoa powder crystallizes into its stable form. Next, the free-flowing powder is passed through sieves and over magnets prior to packing in paper bags or in bulk containers.

Cocoa butter:- The cocoa butter from the presses is filtered and stored. Upon request, the butter can be partly or wholly deodorized. Delivery of the various types of cocoa butter can be either in liquid form or in solid form (packed in plastic-lined cardboard boxes).

Yield:- 650 to 1300 Kg/dry beans.

Important pest of cocoa

1. Mealy bugs (*Planococcus liiacinus* Ckll. and *P. citri* Risso): It is the most important insect pest of cocoa in India. The adult females and young ones feed on the tender shoots, flower cushions, flowers, cherelles and pods by sucking the sap.
2. Tea mosquitoes (*Helopeltis anlonil*).
3. Stem borer (*Zeuzera colleae* Nletn.)
4. Aphids (*Toxoptera auranlii* B de F.)
5. Stem girdler (*Glenea* sp.)
6. Leaf eating caterpillars (*Lympantria* sp.)
7. Leaf eating beetles (*Mylloceros* sp. and *Popillia* sp.)

Important diseases of cocoa

1. Black pod disease (*Phytophthora palmivora*):- It occurs in all the cocoa growing areas in South India during the South-West monsoon period with the maximum incidence in July-August.
2. Canker (*P. palmivora*).
3. Charcoal pod rot (*Botryodlplodia theobromae*).
4. Pink disease (*Pellicularia salmonicolor*).
5. White thread blight (*Marasmius scandens*).
6. Vascular streak die-back (*Oncobasjdium theobromae*).

Chapter 5

Coconut

Botanical Name : *Cocus nucifera* Linn.

Family : Palmae

Origin : South East Asia and Malaysia.

Chromosome No : 2n = 32

Type of fruit : Drupe.

Type of Inflorescence : Spadix

Useable Part : Kernel

- Coconut derives from the Spanish and Portuguese word coco, meaning "a grin", "a monkey face" respectively, since there is a slight resemblance to a human face or a monkey head because of the three tiny indents on the hairy shell of the fruit.
- Coconut known as Kalpavriksha, Tree of Heaven, Tree of Abundance, Tree of Life, King of Palms, Natures Super Market Gods, Gift to Mankind.
- India ranks third on world in coconut area and production.
- The shares of coconut growing countries in production are: Indonesia (25.7%), Philippines (23.2%), India (23%), Sri Lanka (4.4%), others (13.7%).
- India having the highest productivity of 7779 nuts per hectare as against 3630 nuts per hectare in Indonesia and 3859 nuts per hectare in Philippines, the per capita annual availability of coconut estimated to have been 10 nuts only which is quite low compared to 222 of Philippines, 145 of Sri Lanka and 55 nuts of Indonesia.
- Kerala is leading producer state with 766.00 thousand ha area, 3973.90 thousand metric ton production and 5188 Kilogram/ha Productivity.

- Lakshadweep leading state in productivity of coconut (18731 Kilogram/ ha).
- Four southern states put together account for 92% of the total production in the country (Kerala 45.22%, Tamil Nadu 26.56%, Karnataka 10.85%, Andhra Pradesh 8.93% and other states 8.44%).
- Philippines, Indonesia, India, Sri Lanka, Thailand and Malaysia together accounted for 80.65 per cent of the total area under coconut cultivation and about 82 per cent of world production.
- Coconut Development Board established in 1981.
- The most important and economically valuable produce of coconut palm is its fruit popularly known as 'nut'.
- It is madeup of an outer exocarp, a thick fibrous fruit coat known as husk; underneath lies the hard protective endocarp or shell. Lining the shell is a white albuminous endosperm or 'coconut meat' and the inner cavity is filled with a clear sweet refreshing liquid called 'coconut water'.
- The kernel of a matured nut is the most precious product used for edible purpose.
- The dried kernel or copra is the richest source of edible oil and a by-product coconut oil cake, a source of vegetable protein used as an ingredient for livestock feed.
- The shell as such is used for fuel purpose, shell gasifies as an alternate source of heat Coconut fruit energy, making handicrafts, ice-cream cups and other commercial products like shell powder, shell charcoal and activated carbon.
- The pH of coconut water is 4.5.
- The husk yields fibers, which is converted into coir and coir products viz., coir carpets, coir geo-textile, coir composite, coir safety belts, coir boards, coir asbestos and coir pith. Coir pith a secondary by product obtained during defibering process is used as soil conditioner and mending all types of soils.
- Economic life of the coconut palm is about 60 years.
- The coconut palm requires a warm climate without greater diurnal variation of temperature.
- The ideal mean temperature is usually at 27°C with relative humidity more than 60%.

- The palm grows best under a well distributed rainfall between 1,300 mm and 2,300 mm, throughout the year.
- The palm requires plenty of sunlight and does not flourish in cloudy regions. The young palms do grow under shades of old palm during replanting but requires sufficient room for roots and light for growth. Therefore, it makes desirable to remove old palms when palm trees attain the age of eighth year. The growth of palms towards the light or their heliotropism could be noticed on the seashore plantations.
- On sandy soils along the sea coast good for coconut, but also grow on various types of soil, provided there is free-drainage and allow unrestricted root development and aeration, viz. Alluvial soils (Godavari delta), Lateritic soils (Red loam).
- Suitable pH ranging from 5.2 to 8.0.

Coconut Cultivars

Tall Variety

1. **West Coast Tall (WCT):** Extensively cultivated in the West Coast region. Regular bearer producing 12 inflorescences per year. Start bearing in about 6-7 years. Average annual yield 40 to 100 nuts per tree. Average copra content 176 g. Oil content 68 per cent. Recommended for cultivation in coastal region of Kerala and Karnataka.
2. **East Coast Tall (ECT):-** This is grown extensively in the East Coast of India. The palm takes 6-8 years to start bearing. The average annual yield is 70 nuts per palm. Average copra weight of 125 g and 64 per cent oil.
3. **Tipton tall:-** Popular tall cultivars of Karnataka. Average yield is 86 nuts per palm per year. Average copra weight 178 g and oil content of 68 per cent.
4. **Benaulim tall (Pratap):-** Mostly found in Goa, Konkan region of costal Maharashtra. Starts bearing in about 7-8 years. Average annual yield is 150 nuts per palm with copra weight 152 g and 64 percent oil content. The variety is cultivated by name Pratap in Maharashtra state.
5. **Lakshadweep ordinary (Chandra Kalpa):-** The variety is indigenous to Lakshadweep Islands. The palms are good source for tapping toddy. Average yield of 100 nuts per palm, copra weight 176 g with 72 per cent oil content. The variety is released as Chandra Kalpa by CPCRI in 1985 for cultivation in Kerala, Karnataka and Andhra Pradesh.

6. **Andaman ordinary:-** The variety is largely grown in Andaman & Nicobar Islands. The nuts are fairly large in size with average yield of 94 nuts per palm per annum with copra weight 169 g and 66 percent oil content.
7. **Sevvelanir:-** This cultivar found in Pondicherry. The mean annual yield is 40 to 45 nuts per palm.
8. **Kamrupa:-** The variety was originally known as Assam Green Tall, a selection from the local germplasms, one of the most promising cultivars of this region. It has high yielding capacity (106 nuts per palm per year). Copra oil content 64.50 per cent. Tolerant against major coconut pests and diseases commonly affecting coconut palm.
9. **Philippines ordinary:-** It is a exotic cultivar from Philippines. Annual average yield is 110 nuts, with copra weight 183 g and oil content 66 percent.
10. **Philippines laguna:-** Annual average yield 88 nuts, per palm, with copra weight 258.9 g and oil content 66.5 percent.
11. **Fiji longtongan:-** Annual yield of 104 nuts, per palm, copra weight 210.5 g per and oil content 66 percent.
12. **Fiji tall:-** Annual yield of 106 nuts, per palm, with copra weight 199.1 g and oil content 65.2 percent.
13. **S.S. green**:- Annual yield of 108 nuts, per palm, with copra weight 186 g and oil content 67 percent.
14. **Sanramon:-** Annual yield of 64 nuts, per palm, with copra weight 349.6 g and oil content 68 percent.

Dwarf Cultivars

1. **Chowghat orange dwarf:-** This variety is found particularly in Chavakkad area of Trissur district in Kerala. It is also known as Gaurigathram or Chenthengu and Kenthali in Karnataka. Starts bearing from 3-4 years. Average annual yield is 65 nuts per palm, average copra weight 150g. with 66 percent oil content. The variety is ideal for Tender coconut water purpose. The variety is recommended for cultivation in Kerala and Karnataka.
2. Chowghat Green dwarf
3. Malayan Yellow dwarf
4. Malayan Orange dwarf

5. Malayan Green dwarf
6. Ganga bondam
7. Kentholi Orange dwarf
8. Chitta gangapani
9. Udha gangapani

Hybrid Cultivar

1. **Chandra shankara**:- Cross between COD x WCT. Annual yield 222 per palm with copra weight 215g and oil content 68 per cent. The variety is recommended for Kerala.
2. **Kera shankara:-** Cross between WCT x COD. Annual yield 211 per palm with copra weight 187g and oil content 68 per cent. The variety is recommended for Kerala, coastal Maharashtra and coastal Andhra Pradesh.
3. **Chandra laksha**:- Cross between LCOT x COD. Annual yield 171 per palm with copra weight 195g and oil content 69 per cent. The variety is recommended for Kerala.
4. **Laksha ganga:-** Cross between LCOT x GBGD. Annual yield 186 per palm with copra weight 186g and oil content 70 per cent. The variety is recommended for Kerala.
5. **Ananda Ganga:-** Cross between AODT x GBGD. Annual yield 95 per palm with copra weight 216g and oil content 68 per cent. The variety is recommended for Kerala.
6. **Keraganga**:- Cross between WCT x GBGD. Annual yield 100 per palm with copra weight 201g and oil content 69 per cent. The variety is recommended for Kerala.
7. **Kerasree**:- Cross between WCT x MYD. Annual yield 130 per palm with copra weight 216g and oil content 66 per cent. The variety is recommended for Kerala.
8. **Godavari ganga :-** Cross between ECT x GBGD. Annual yield 140 per palm with copra weight 140g and oil content 68 per cent. The variety is recommended for Andhra Pradesh.
9. **Kera sowbhagya:-** Cross between WCT x SSA. Annual yield 116 per palm with copra weight 116g and oil content 65 per cent. The variety is recommended for Kerala.

10. **VHC-1:-** Cross between ECT x CDG. Annual yield 98 per palm with copra weight 135g and oil content 70 per cent. The variety is recommended for Tamil Nadu.

11. **VHC-2:-** Cross between ECT x MYD. Annual yield 107 per palm with copra weight 107g and oil content 69 per cent. The variety is recommended for Tamil Nadu.

12. **VH-3:-** Cross between ECT x MOD. Annual yield 156 per palm with copra weight 156g and oil content 68per cent. The variety is recommended for Tamil Nadu.

- Pit size for planting:- 1x1x1 m.
- Planting distance: - 7.5x7.5.
- Propagation method:- Seed.

Mother Palm Selection for Seed

- A good regular bearing mother palm produces on an average one leaf and an inflorescence in its axil every month. So, there will be twelve bunches of varying stages of maturity at any one time with strong bunch stalks. Avoid trees producing habitually barren nuts.
- Straight stout trunk with even growth and closely spaced leaf scars.
- Spherical or semi spherical crown.
- High rate of leaf (more than 30 fully opened leaves) and spathe production (12 inflorescences).
- Short and stout petiole and wide leaf base firmly attached to the stem,
- Short and stout inflorescence stalk with bunches, preferably resting on the leaf petioles of the lower whorl and more number of female flowers (25 or more).
- The age of the palm chosen be middle age i.e., from 25 to 40 years. Even trees with 15 years age can be selected, if it is high yielding and has stabilized yield.(eg, Chowghat dwarf). Avoid palms that are above 60 years.
- High yielding mother palms giving not less than 100 nuts/palm/annum under irrigated condition (70-80 nuts/annum under rainfed conditions) should be chosen for collecting seed nuts.
- Husked nuts should weigh not less than 600 g.

- Mean copra content of 150 g per nut or more.
- Free from pest and diseases.

Selection of Seed Nuts:- The mature nuts are harvested when atleast one nut in the oldest bunch starts becoming dry. In Tall, it takes 11-12 months to become a matured seed nut whereas in dwarfs, nuts will mature in 10-11 months after emergence of the inflorescence. In general dwarf palms are to be harvested once in four weeks and tall palms are to be harvested once in six to eight weeks. They produce a resonant and ringing sound when hit with the harvesting knife or tapped by finger indicating that the husk is dry. Immature nuts will produce a dull sound. Harvest the bunches intended for seed nut by lowering them to the ground using a rope to avoid injury to seed nuts when palms are tall and ground is hard. The seed nuts should be medium sized, round or oblong in shape.

- Vigorous seedlings which are one year old, having minimum of six leaves and girth of 10 cm at the collar level should be selected for planting in the main field (9-12 month old).
- Planting the seedlings during May with the onset of pre-monsoon rain is ideal.

Manures and Fertilizers

Particulars	Year 1	Year 2	Year 3	Year 4 onwards
FYM (kg/palm)	40	20	25	30
N (g/palm)	50	160	330	500
P (g/palm)	40	120	240	320
K (g/palm)	135	400	810	1200

- Fertilizers may be applied in two split doses during May-June and September-October for the rain fed palms and in four or more equal splits for irrigated palms avoiding the heavy rainfall period.
- To maximize the utilization of soil and sunlight in the coconut garden, intercropping can be adopted with a variety of crops like pineapple, banana, elephant foot yam, groundnut, chillies, sweet potato, and tapioca etc. up to 8-10 years.
- During 10-22 years of age of the palms, crop like colocasia which can tolerate shade can be cultivated. In older plantations, perennials like cocoa, pepper, cinnamon, clove and nutmeg can be grown as mixed crops along with intercrops.

- Coconuts become mature in about 12 months after the opening of the spathe. It is the ripe coconut which is the source of major coconut products.
- Nuts which are eleven months old give fiber of good quality and can be harvested in the tracts where green husks are required for the manufacture of coir fiber.
- The nuts must be fully matured for good quality ball copra and seed purpose. Besides, coconut oil, desiccated coconut is also an important commercial product for which fully ripped nuts are essential. Usually, 11 to 12 months old nuts are harvested at varying intervals in about 6 to 10 times in a year depending upon the yield of palms.

Important insect of coconut

1. Rhinoceros beetle (*Oryctes rhinoceros* L.) is attack on Spindle, leaves, Spathe, destruction of inflorescence.
2. Red Plam Weevil (*Rhynchophorus ferrugineus* Fab.) is attack on Terminal bud, upper portion of trunk spindle, tender leaves.
3. Leaf eating Caterpillar (*Opisina arenosella* Wik.) is attack on leaves, particularly lower leaves surface during summer months from February to June.
4. White grub (*Leucopholis coneophora* Burm.) is attack on Roots affected palms develop sickly pale yellow leaves, exhibits Button shedding and tapering of crown region.
5. Eriophyid mite (*Aceria guerreronis*) infests the soft coconut by sucking sap from tissues of buttons. Triangular patches close to perianth due to sucking sap. The mite causes death of meristematic tissues resulting formation of brownish patches on the surface of nuts causing warting and longitudinal fissures. Prevents growth of nuts, reduction in shell and kernel. Affects quality and quantity of fiber content of husk.

Important disease of coconut

S.No.	Disease	Causal Organism	Parts affected
1.	Bud-rot or Fruit rot (Mahali)	*Phytophthora palmivora* (Fungus)	Yellowing of younger leaves Withering of spindle (0.1 to 10% incidences).
2.	Basal stem rot or Thanjavur wilt	*Ganoderma lucidum* *G.aplanatum* (Fungus)	Decay of roots system, flaccidity of spindle leaves browning of outer leaves. (2.6 to 13 % incidences).
3.	Stem bleeding	*Thielaviopsis paradoxa* (Fungus)	8 – 9 per cent exudation of the dark reddish brown liquid from the longitudinal cracks in bark at the base of the trunk.
4.	Leaf rot	*Colletotrichum gloeospori -oides,Exserohilum rostra-, fun Fusarium solani* (Fungus)	Blackening and shriveling of distal ends of leaflets in central spindle and younger leaves.
5.	Leaf blight or Greyleaf spot	*Pestalotia palmarum* (Fungus)	Necrosis of mature leaves causes extensive leaf blight leading to complete drying of leaf blade and shrivels off.
6.	Root (Wilt) Disease	Phytoplasma Transmitted by Vector *Stephanotis typicus* bug and plant hopper *proutista moesta*.	Flaccidity bending of leaves, yellowing and marginal necrosis of leaflets. Smaller nuts and thin kernel.
7.	Tatipaka disease	Phytoplasma	
Physiological disorder			
8.	Crown Choke Disease	Deficiency of boron	Deformed leaves severe tip necrosis choked appearance of frond.

Coconut by product

1. **Tender coconut:** It take 12 months period between the emergence of inflorescence and complete maturity of the fruits.

2. **Snow ball tender coconut:** The snow tender kernel of tender Coconut is a delicious dessert. Traditionally, the kernel of tender coconut is sometimes consumed after drinking the water by the consumer or else the ball is thrown away.

3. **Egg coconuts:** Eggs coconut referred to a form of product derived from tender coconut where a whole around soft kernel is pulled out from its shell with tender water intact in it.

4. **Pre packed tender coconut water**

5. **Bottled coconut water:** Matured coconut water can be marketed as natural soft drink if preserved and packed in bottles as non-carbonated beverage. The pH of the coconut water is adjusted to 4.2 with citric acid and 0.1–0.15 per cent sodium citrate as the salt reduces the bitter taste that may develop on addition of citric acid to stabilize the acidic pH.

6. **Coconut vinegar:** Coconut water can be converted into vinegar by using vinegar generator assembly which comprises a feed vat, an acidifier and a receiving vat for collection of vinegar. The mature coconut water consisting of about 3 per cent sugar content is concentrated to 10 per cent level by fortifying with sugar.

7. **Nata–de–Coco:** It is a gelatinous dessert delicacy produced by action of bacteria *Acetobacter xylinium* in culture medium of matured coconut water. The product is used as a dessert and as an ingredient in ice-cream and fruit mixes.

8. **Coconut milk:** Coconut milk is a product extracted from the endosperm or kernel of coconut and constitutes into an emulsion of coconut oil in water, stabilized by proteins and probably by some iron found in oil water interface.

9. **Sweet coconut chips:** The product is developed from the matured coconut kernel after removing the moisture content of the kernel through osmotic dehydration. The dehydrated coconut chip in ready-to-eat form can be used as snacks.

10. **Virgin coconut oil:** The virgin coconut oil is obtained from the fresh coconut meat (kernel) of coconut by chemical or natural means with or without the application of heat, which does not alter the oil and its properties. Virgin coconut oil is edible in its natural state immediately after expulsion and filtration. It is a pure form of coconut oil, crystal white in colour, with natural vitamin E (Tocopherol) contents. It has a mild scent of fresh coconut.

11. **Coir:** Coir is one of the important natural, golden colour, hard fibers extracted from the husk or fibrous mass (mascara) covering the coconut. India is the largest producer of brown coir fiber.

12. **Coir pith :** Coir pith commonly known as coconut peat is the by-products of coir industry. Coir pith is converted into briquettes for the purpose of easy transportation. It is easily composed to be used as an organic manure and soil ameliorant.

13. **Neera:** The vascular sap collected from immature unopened coconut inflorescence is popularly known as "Neera" in fresh form. It is a sugar containing juice, a delicious health drink and a rich source of sugars, minerals and vitamins. It is sweet and oyster white in colour and translucent. It is tapped from the coconut inflorescence filtered, pasteurized, and bio preservatives added to preserve the product. Treated Neera can be preserved in cans up to 2 months at room temperature.

14. **Coconut palm jaggery:** The strained unfermented coconut sap is boiled, crystallized and transferred into suitable moulds to prepare jaggery. The semisolid jaggery solidifies gradually by the cooling effect of the atmosphere into a crystallized hard substance. The recovery of jaggery from syrup is 15%. It is used as a sweetening agent for the preparation of dishes and is superior to cane jaggery.

15. **Coconut palm sugar:** The coconut palm syrup or jaggery can be crystallized to produce fine granules of sugar.

16. **Coconut toddy:** Toddy on fermentation becomes an alcoholic drink. Arrack and vinegar are also manufactured from coconut toddy. In Goa commercial arrack obtained by distillation of coconut toddy is known as coconut fenny.

17. **Coconut oil cake:** Coconut cake is the residue left after the extraction of oil from copra which is mainly used as a cattle feed. Coconut cake contains 4-5 per cent oil which is extracted by solvent extraction process.

Chapter 6

Tea

Chinese big leaf tea (*Camellia sinensis* var. *macrophylla*).

Chinese small leaf tea (*Camellia sinensis* var. *bohea*)

Shan tea (*Camellia sinensis* var. *Shan*)

Indo-China or Cambod tea (*Camellia assamica* subspecies *lasiocalyx*)

Indian tea (*Camellia sinensis* var. *assamica*)

Family: Theaceae

Chromosome No.: 2n = 2x = 30

Origin: China (East and South Asia)

Useful Part: Bud and Leaves

	Assam Jats (*Camellia assamica*)		**China Jats (*Camellia sinensis*)**
1.	It is tree.	1.	It is shrub.
2.	Few robust branches.	2.	Branches abundant and whippy.
3.	Large, glossy leaves.	3.	Small leathery leaves.
4.	Light to medium green.	4.	Dark green colour.
5.	High yield and medium quality.	5.	Low yield but good quality.
6.	Susceptible to drought and frost.	6.	Hardy and resistant.
7.	Sparse flowering.	7.	Profuse flowering.

- In its wild state the tea plant is a small tree or shrub. It was first named *Thea sinensis*, and is now known as *Camellia sinensis*.
- The tea was first used in China as a medicinal drink, and later became a popular beverage.
- 2737 BC - An early emperor in China , Shen Nung who was a skilled ruler and creative scientist accidentally discovered tea.

- 618–906 AD - Tang Dynasty powdered tea became the fashion of the time. Nobility made it a popular pastime. Caravans carried tea on the Silk Road, trading with India, Turkey and Russia.
- 805 - The Buddhist monk Saicho brought tea seeds to Japan from China.
- 1657 - Tea was first sold in England at Garway's Coffee House in London. The East India Company operated out of several sites in the City of London.
- 1706 - The first auction dedicated to tea took place in Craven House, which became known as East India House.
- Britain is a great consumer in average an individual uses 4.4 kg of tea annually.
- Caffeine constitutes 2.5-5.5% of dry matter of tea leaves. It is importance in the taste of tea. Theobromine (0.07-0.17%) & theophylline (0.002-0.013%) are also present that are responsible for colour and flavor.
- The average annual per capita consumption in the world is 0.75 kg.
- In India per capita consumption is 0.3 kg.
- India is second largest producer of Tea.
- Major tea-growing states are Assam (53%), West Bengal (23.9%), Tamil Nadu (11.3%) and Kerala (8.44%).
- According to the Planning Commission Deputy Chairman, Montek Singh Ahluwalia, recognise tea as the "National Drink".
- The constituent responsible for colouring of tea are Theaflavin and Thearufigens.
- Tea an evergreen medium sized shrubs (4-6 ft.) that grows mainly in tropical and subtropical climates.
- Flowers are solitary in the axils of the leaves. The flowers are white, middle yellow & 1 ½ inches in diameter and are fragrant.
- Tea is propagated by cutting (semi-hard stems) or seed.
- Tea requires a moderately hot and humid climate.
- Rainfall: The average annual rainfall of 2000 mm is required.
- The monthly average maximum temperature varies between 22 and 28°C during October–March, but it may rise up to 37°C between April and September. The minimum temperature usually does not go below 7°C.

- Generally, tea thrives within latitude 43^0 North and 27^0 South.
- Tea grows well on high land well drained soils having a good depth, acidic pH in the range 4.5 to 5.5 and more than 2% organic matter.
- Planting method:- Single or double hedge style.
- Planting distance

Style	Spacing	Population/ha
Up and Down	1.2 x 1.2 m	6,800
Contour planting single hedge	1.2 x 0.75m	10,800
Contour planting double hedge	1.35 x 0.75 x 0.75	13,200

Pruning carried out occasionally is a critical operation.

Type of pruning	Pruning height (cm)	Season	Remarks
Rejuvenation pruning	20cm in China Jat 30cm in Assam Jat	April-May	Done in old bushes affected with canker and wood rot to invigorate the new healthy branches. Not done regularly.
Hard Pruning	30–45cm	April-May	First formative pruning done to young tea.
Medium Pruning	45 and 60cm	August-September	Normal pruning.
Light pruning	60-70cm	August-September	Normal pruning.
Skiffing	70-75cm	August-September	Mainly to postpone pruning and to encourage better frame development.

- Tea is a semi-shade loving plant.
- Dense shade increase the incidence of blister blight disease in south India.
- In north-east, trees recommended for shade are *Albizia chinensis, Albizia odoratissima, Albizia lebbek, Acacia lentiartaris, Derris robusta* and *Dalbergia sericea.*
- In south India, *Grevillea robusta* is the only tree recommended for permanent shade.
- Paraquat dichloride, Paraquat glyphosate and 2, 4-D are used as a weedicide in tea garden.
- Sprinkler irrigation is the most common method in tea garden.

Varieties of tea

S.No.	Variety/clone	Remarks
1.	Darjeeling	Darjeeling tea ranks among the finest and most expensive varieties in the world.
2.	Dooars	Tea growing region in Northern India, situated west of Assam, strong flavour reminiscent of Assam tea.
3.	Dragon Pearl	Very rare and special tea with a hand-rolled, green leaf and Silver Tips (known also as Silver Needles).
4.	Hazelbank	This small, picturesque garden in the Indian region of Assam produces one of the best tea in the world. It is named after Hazel, daughter of the famous civil servant Dr. Mead. The fullness, size and colour of Hazelbank makes it a favorite among tea merchants.
5.	Jasmine tea	Probably the best known flavoured Chinese tea. Perfumed exclusively with tender white jasmine flowers. These lose their scent within 20 hours of picking, and therefore have to be added to the tea immediately to give it the desired aroma.
6.	Snow Buds	Green tea with white tips and a pleasant 'White Tea character', honey-coloured liquor. Mild, smooth taste.
7.	UPASI 1(Ever green)	Hardy, and good quality.
8.	UPASI-2 (Jayaram)	It is an average yielding clone, tolerant to drought and wind.
9.	UPASI-3 (Sundaram)	Very high yielding and good quality clone.
10.	UPASI-6 (Brooklands)	Suitable for low elevations.
11.	UPASI-8 (Golconda)	High yielding and suitable for all elevation.
12.	UPASI-9 (Athrey)	High yielding, tolerant to drought and can grow in slightly high pH soil.
13.	UPASI-10 (Pandian)	It is a high yielding clone, resistant to drought and wind. It is suitable for high elevation.
14.	UPASI-14 (Singara)	Suited to higher elevations, High yield clone
15.	UPASI 15 (Spring field)	Flushes throughout the year.
16.	UPASI 17 (Swarna)	Flourishing well at mid and high elevations.
17.	UPASI 28	Cross between UPASI 10 x TRI2025. Biclonal, Good strength and high quality.

Other cultivars and clones

White tea, Java, Dimbula, English breakfast, Genmaicha, Green monkey, Green mu dan, Green pekoe, Gunpowder, Gyokuro, Highgrown, Kwai flower, Lapsang souchong, Matcha, Muscatel, Pettiagalla, Silver sprout, Tea manufacture, UPASI-4, UPASI-5, UPASI-7, UPASI-11, UPASI-12, UPASI-13, UPASI-16, UPASI-18, UPASI-19, UPASI-20, UPASI-21, UPASI-22, UPASI-23, UPASI-24, UPASI-25, UPASI-26 UPASI-27, AK 1, ATK 1, B-1-101, BSB 1, BSS I,

BSS II, BSS III, CH-1 CR6017, SA 6, SMP 1, TES 34, TRF 1, TRI2024 (Sri Lanka), TRI2025 (Sri Lanka), TRI2026, TRI2043, TTL 1, TTL2, TTL 3, W-2, TRF-2 and CR – 6017.

Manure and Fertilizer

Year of application	Total weight kg/ha/annum			No. of applications
	N	P	K	
Ist year	180	80	270	5
IInd year	240	100	360	6
IIIrd year	300	100	450	6
IVth year onwards	300	100	300	6

Leaf harvest and transport: Harvesting is most important cultural operation which influences yield and share 70% of total cost of cultivation.

- The first harvest is obtained after 4-5 years.
- The shrub can be used for 60-70 years.
- Harvesting season depends upon the region and climate and lasts for 8-9 months per year.
- Leaves can be plucked at intervals of 6-9 days.
- Generally tea is harvested at a bud and 2-3 leaf stage.
- Most common and important stage is a bud and 2 leaf stage.
- An average plucking standard should have ideally about 75% fine leaf by weight comprising bud and two leaves.
- When the crop shoot is harvested, generally the bud on the axil of the top most leaf of the stem develops into a new crop shoot. During the development of this bud, first 2 scale leaves (cataphylls) or '*Janam*s' appear. Above the scale leaves or *Janam*s is the smooth 'fish leaf'. This is followed by a series of 3 or 4 normal leaves, the first of which is called the mother leaf.
- The introduction of skiffing in place of pruning in North-eastern India has helped to increase the productivity.
- Skiffed bushes are more prone to drought and they get heavily infested by pests such as red spider mite and tea mosquito bug.
- The first plucking of recovering bushes is called tipping. The objective of

tipping is to establish a level plucking surface, to provide adequate maintenance foliage and for the quick production of secondary branches.

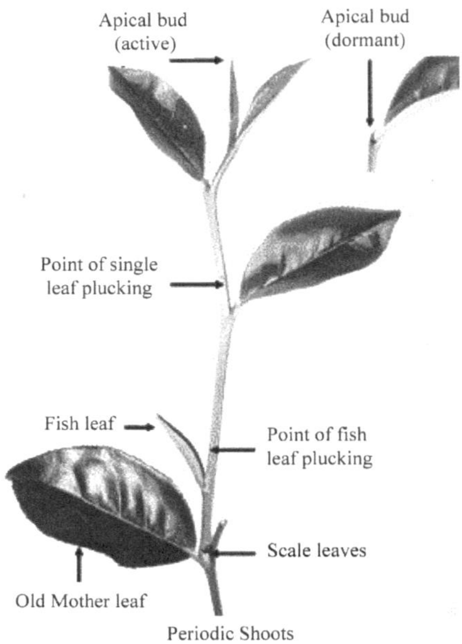

- Shoots may be plucked above the *Janam*, fish leaf or mother leaf. These systems of plucking are known as *Janam* plucking, fish leaf plucking and mother leaf plucking, respectively.
- In North-eastern India, *Janam* plucking is the most common practice.
- Tea shoots are plucked to the level (*Janam* or fish leaf) during April–December and a new tier of leaf is added to the bush by resorting to mother leaf plucking from January to March. Plucking round is the number of days between 2 successive plucking.
- The stems produced by a bush immediately following pruning are produced from buds that had remained dormant for some time. These stems are distinct from the normal flush growth as they do not develop into *Banji* and continue to produce leaves at fairly regular intervals. They are known as periodic shoots.

Processing of Tea

Withering - The first processing step in the factory and is a process in which freshly plucked leaf is conditioned physically, as well as, chemically for subsequent processing stages. The main objective of withering is to remove about 15–20% of moisture from the leaves.

Rolling - The withered leaf is charged in the rollers which vary in size. Various enzymes and their substrates come in contact and enzymatic oxidation starts at this stage. The green colour of leaf starts diminishing and brown coppery colour of leaf begins to appear. There are two methods:-

- **Orthodox process:** In the orthodox process of manufacture disruption of the leaf cells is carried out in a Rolling Table. During rolling juice from the leaf is wrung out and the leaf is also twisted and broken into smaller particles.

- **The CTC (Crush, Tear, Curl) Process:** The CTC process achieves the three actions of Crushing, Tearing and Curling in the same machine at one go. Instead of working on large volumes of leaf, the CTC takes a fast, thin, but steady stream of leaf to pave the way to continuous processing in place of the batch mode of orthodox rollers.

Fermentation (Oxidation): The process objective is to allow intimate contact of the catechins with the respective enzymes, which oxidize these catechins in presence of oxygen. Oxidation of catechin /polyphenols by the enzyme polyphenol oxidase is the main reaction of the process that imparts 'tea character' to the black tea.

Drying: Fermented leaf is loaded onto the dryer, which comprises perforated moving trays. Thickness of spread, speed of trays and volume of air blown through are regulated to achieve the correct drying.

Sorting: The bulk tea is sorted on mechanically oscillated sieves. The sieve size decreases gradually from top to bottom.

- On an average 100 kg of fresh leaf produces 22.5 kg of dried tea containing residual 3% moisture.

Types of Tea

- **Black tea:** Leaves are fully fermented.
- **Oolong tea:** Leaves are partially fermented. The same process applied to black tea applies, but the fermentation time is substantially shortened.
- **Green tea:** Leaves are not fermented at all. The leaves are heated immediately after picking to prevent fermentation. The leaves are then rolled, dried, and packed. Enzyme killed by placing in boiling water.
- **White tea:** Steamed and dried almost immediately after harvesting.
- **Scented tea:** Scented tea, which is very popular in Northern China, in fact is a mixture of green tea with flower petals of rose, jasmine, orchid and plum through an elaborate process. Among this type, Jasmine tea is common.
- **Silver tips:** All silver tip teas are manufactured only from buds that are picked during a period of 3 to 4 days just before the beginning of the main plucking season. The silver tips brews a fine silvery color infusion with lots of delicate aroma and wonderful taste.

- **Golden tips:** Uses only buds with a golden tint. Brews an outstanding cup full of delicate flavor and wonderful taste. A rare tea only available in this limited quantity every year.

Grade of Tea

S. No.	Grade	Grade abbreviation	Remarks
1.	PEKOE		
2.	PEKOE1	PEKOE one	Same as PEKOE, but smaller in size than PEKOE of any elevation. This replaces the Flowery PEKOE grade. (PEKOE and PEKOE1) will be treated as two grade but for cataloguing purpose treat as one grade).
3.	FBOPF		Similarly in size to BOP, BOPF and must conation tips.
4.	FBOPF1		Larger than BOP, smaller than FBOP with a show of tips.
5.	FBOPF Sp		Similar in size of BOP1 with fair presence of tips.
6.	FBOPF Ex Sp		Small leaf and must have an attractive show of golden or sliver tips with little black leaf.
7.	FBOP Ex Sp1		Leafy and must have an attractive show of golden or sliver tips with little black leaf.
8.	FBOP Ex Sp1		Leafy and must have an attractive show of golden or sliver tips with little black leaf.
9.	OP1	Orange Pekoe One	Long wiry than OP1, but much twisted than OPA.
10.	OP	Orange Pekoe	Less wiry than OP1, but much more twisted than OPA.
11.	OPA	Orange Pekoe	Long bold leaf tea with fair twist.
12.	BP	Broken Pekoe	(Off Grade) Should be choppy, hard leaf.
13.	BOP1A	Broken Orange Pekoe One	(Off Grade) Any flaky leaf without stalk and fiber.
14.	BM		(Off Grade) Mixed flaky leaf tea. Can have more fiber and stlk than BOP1A.
15.	BT		(Off Grade) All mixed teas of varing size with or without stalk than BOP1A.
16.	FNGS		(Off Grade) Flaky leaf of small size. Can contain more fiber than BOPF but Responsibly.
17.	FNGS		(Off Grade) Same as Fanning1 can be more fiber and uneven not as clean as Fannings1.
18.	DUST1		Smaller than BOPF, Grainy even, well made and reasonability clean.

S.No.	Grade	Grade abbreviation	Remarks
19.	DUST		(Off Grade) Similar in size to Dust 1. Could be Flaky and Contains some Fiber.
20.	Sliver Tips		Long tippy leaf, golden in colour, with hardly and black leaf.
21.	Golden Tips		Long tippy leaf, golden colour, with hardly and black leaf.

CTC Tea Grade

S.No.	Grade	Mesh size
1.	Flowery Pekoe	>8 mesh
2.	Pekoe	8-10
3.	BOP	10-12
4.	Pekoe Fanning	12-16
5.	BOP-fanning	12-16
6.	Pekoe Dust (PD)	16-20
7.	Red Dust (RD)	20-30
8.	Super Red Dust (SRD)	40-50
9.	Final Dust	50-60
10.	Superfine Dust	Above 60

Yield:- The average annual yield from small holdings hovers around 2,500 kg of made tea /hectare, and this translates to 10,000 kg of green leaf.

Important Insect and Pest of tea

Tea Mosquito Bug: *Helopeltis theivora*

- **Damaging stage** : Nymphs and Adults
- **Nature of damage**: They suck sap from leaves, buds and tender stems.
- **Management**: Collection and destruction of parts with pest. Spraying with Dimetoate 1.7 ml / ltr.

Tea green leaf hopper: *Emposca flavescens*

- **Damaging stage:** Nymphs and Adults
- **Nature of damage:** They suck sap from leaves
- **Management:** Natural Management: Ladybeetles, Neuropterans, praying mantids feed on leaf hoppers. Tea growing under shade trees tends to

have less leaf hoppers problem. Mulching also helps to increase humidity and therefore reduce leaf hopper populations. Spraying with *Verticillium lecanii* @ 2 g/ltr. Use of light traps. Sprays with Monocrotophos @ 1.25 ml/ltr or Imidacloprid 0.25ml/ltr

Tea tortrix : *Homona coffearia*

Bunch catrerpillar: *Andraca bipunctata*

- **Damaging stage:** Larva
- **Nature of damage:** It first feeds on leaf surface by scraping later feeds on leaf margins and they remain clustered in bunches on branches
- **Management:** Collect and destroy the clustered leaf bunches on branches, Spray with Quinolphos 2 ml/ltr

Tea semilooper: *Biston supprersaria*

- **Damaging stage:** Larva
- **Nature of damage:** It first feeds on leaf surface by scraping later feeds on leaf margins.
- **Management:** Collection and destruction of infested branches. Sprays with Quinolphos 2 ml/ltr

Tea mites

1. Red spider mites
2. Scarlet mite
3. Purple mite
4. Yellow mite

Shot-Hole Borer : *Xyleborus fornicates*

- **Damaging stage:** Adult
- **Nature of damage:** It makes holes on the roots and on the lower part of main trunk. Grubs, upon hatching feed on a fungus *Monocrosporoum embrocium.*

Red stem borer : *Zeuzera coffeae*

Important disease of Tea

- Dieback and wood rot and *Armillaria mellea* root rot
- Blister blight (*Exobasidium vexans* Massee)
- Grey blight (*Pestalotiopsis* spp.)
- Stem canker/basal rot (*Phomopsis theae*)
- Brown blight (*Colletotrichum camellia* Massee)
- Wood rot (*Hypoxylon serpens*) and dieback (*Nectria poria*)
- Root diseases such (*Botryodiplodia theobromae*)
- Brown root disease (*Fomes noxius*)
- Stem end rot (*Thyronectria pseudotrichia*)
- Thorny stem blight (*Aglaospora aculeata*)

Chapter 7

Rubber

Botanical Name: *Hevea brasiliensis*

Family: Euphorbiacace

Origin: Brazil

Chromosome No.: 36

Useful Part: Latex

Cera Rubber: *Manihot glaziovii*

Para Rubber: *Hevea brasiliensis*

Indian Rubber: *Ficus elastic*

Panama Rubber: *Castiolla elastic*

Guayul Rubber: *Parthenium argentatum*

- *Hevea brasiliensis* is mostly used for natural rubber.
- Natural rubber was introduced by Britshe in 1873 at the Botanical Gardens, Calcutta.
- Rubber is a non-wood forest product that has been utilized since ancient times.
- Gestation period- 6-7 years.
- Economic life – around 32 years.

Botany

Rubber is a deciduous tree, typically 30-40 m tall, though usually 15-25 m tall in cultivation, with a leafy crown. The trunk is cylindrical, but frequently swollen towards the base, and the bark is pale to dark brown with a smooth surface and the inner bark pale brown with abundant white or cream coloured latex. The leaves are in spirals and with three leaflets. The flowers are small with no petals, bright or cream-yellow in colour and extremely pungent. They are either male or female but both are found in the same inflorescence. The fruit is an exploding 3-lobed capsule.

- In the wild, trees may grow to over 40 m and live for 100 years, but in plantations they rarely exceed 25 m because growth is reduced by tapping for rubber. Plantation trees are usually replanted after 25-35 years when yields fall to an uneconomic level.
- The milky latex of rubber, produced by a specialized secretory system in the phloem, is the raw material for natural rubber.
- In India Commercial cultivation started in 1902 in Kerala.
- India rank 1st in productivity, 4th in production, and 4th in consumption and 5th in area.
- Ideally rubber grows in the region between 10° North and South of the equator where the rain fall, humidity and temperature are high and the elevation is below 450 meters.
- It requires deep and lateritic fertile soil with an acidic pH of 4.5 to 6.0 and highly deficient in available phosphorous.
- Tropical climate with annual rainfall of 2000–4500 mm is suited for cultivation. Minimum and maximum temperature should be ranged from 25 to 34°C with 80 % relative humidity is ideal for cultivation.
- Regions prone to heavy winds should be avoided.
- In India rubber is traditionally grown in Kerala and Kanyakumari district of Tamil Nadu
- It is a sturdy tree that can withstand some variation in climate. Climatic variations will be reflected in the growth and yield.

Method of propagation

- Green budding, brown budding and crown budding.
- In India, the peak season for harvesting Rubber is from October to January while the lean period is during monsoon.
- Rubber plants generally have 32 years of economic life but they may live up to 100 years or even more than that.

- The plantation would start its yield from 6th year onwards. The natural rubber produced is processed to convert into a storable and marketable form.
- The basic property of rubber is that it comes back to its original shape if it is twisted or stretched but if heat is applied to the rubber, it won't return to its original shape easily.
- The rubber industry produces wide range products like auto tyre, auto tubes, automobile parts, footwear, belts, cables and wires, battery boxes etc. Block rubber, Preserved latex, Crepes and sheets are some forms in which rubber is produced and used.
- India is the fourth largest producer of Natural rubber, after Thailand, Indonesia and Malaysia.
- Thailand is the world's largest producer of Natural rubber. It produced 3,394 thousand tonnes in 2011. Indonesia is the second largest producer contributing around 27% of the world's total natural rubber production. Malaysia produced 996 thousand tonnes and occupies the third rank in terms of production. India is the fourth largest producer of natural rubber in the world. In the year 2011, India produced 890 thousand tonnes of natural rubber, up from 851 thousand tonnes in the previous year. India contributes 8% of the world natural rubber production.
- Other major producers include Vietnam and China contributing 7% each to the world production.
- As mentioned earlier, Rubber plantations give yield only after 6 years of being planted and are productive for about 25 years.
- Indonesia, the second largest producer has 3,435 thousand hectares under rubber. While, Thailand which is the largest producer has 2,756 thousand hectares planted with rubber. This implies that Thailand has greater yield than Indonesia.
- Malaysia and China have 1,237 and 932 thousand hectares under rubber respectively. India has around 687 thousand hectares under rubber plantations.
- As per the Rubber Board, India now occupies the first rank in terms of productivity at 1,819 kg/ha.
- China being the fastest growing economy of the world is also the world's largest natural consuming country. At 3,603 thousand tonnes it consumes 33% of the total rubber consumed over the world.

- USA has replaced India this year, being the second largest consumer of rubber at 1,029 thousand tonnes, followed by India at 958 thousand tones.
- India's Natural rubber production has been increasing steadily over the past decade.The production grew at the rate of 4.8% in the year 2011-12. India produced 904 thousand tonnes in 2011-12 compared to 862 thousand tonnes in 2010-11.
- The rubber growing regions in India can be classified under two major zones, traditional and non-traditional, on the basis of agro-climatic conditions.
    - **Traditional Regions:** Rubber cultivation in India has been traditionally confined to hinterlands of southwest coast, mainly in Kanyakumari district of Tamil Nadu and Kerala.
    - Kerala and Tamil Nadu together constitute the traditional rubber growing regions in the country.
    - Kerala alone contributes 89% of the total rubber produced in India and an area of 534,228 ha under rubber plantations.
    - Tamil Nadu contributes another 3% of the total natural rubber production.
    - **Non-Traditional Regions:** These are hinterlands of coastal Karnataka, Goa, Konkan Region of Maharashtra, hinterlands of coastal Andhra Pradesh and Orissa, the Northeastern states, Andaman and Nicobar Islands etc, where rubber is now being grown

State	Production (Tonnes)	Area (Ha)	Share in total production
Kerala	770,580	534228	89%
NE Region	41,200	113685	5%
Tamil Nadu	25,160	19767	3%
Karnataka	23,705	38110	3%

- In recent years among nonTraditional region, Tripura, has become one of the most thrust areas for Rubber growing because of its well acceptance worldwide. In fact, Tripura was declared the "Second Rubber Capital of India" by the Rubber Board.

Rubber cultivar

S.No.	Cultivar	Character
1.	RRII 105	A clone evolved by the Rubber Research Institute of India and currently enjoying maximum popularity in the country. Parents are Tjir1 and Gl 1. Yield is very good. The average commercial yield obtained over 22 years of tapping in various estates is 1712 kg/ha/yr. Latex is white and DRC high. This clone has tolerance to abnormal leaf fall disease. Highly susceptible to pink disease and Corynespora leaf disease.
2.	PB 260	A hybrid clone developed by the Prang Besar Estates Ltd.in Malaysia. Parents are PB 5/51 and PB 49. Yield is high. Mean yield over ten years of tapping is 1884 kg/ha/yr. Average yield from large scale trials over 10 years is 63g/tap/tree. The clone has moderate tolerance to pink, powdery mildew and abnormal leaf fall. High incidence of tapping panel dryness.
3.	RRII414	The parents of this clone are RRII 105 and RRIC 100. Moderately tolerant to Pink disease and abnormal leaf fall. Incidence of *Corynespora leaf* fall is low and powdery mildew is high.
4.	RRII 430	Parentage comprises RRII 105 and RRIC 100. Tolerant to pink disease, abnormal leaf fall and *Corynespora* leaf fall, but susceptible to powdery mildew.
5.	RRII 417	The parentage of this clone comprises RRll 105 and RRIC 100. Susceptible to powdery mildew and moderately tolerant to pink disease, abnormal leaf fall and *Corynespora* leaf fall.
6.	RRII 422	Parents of this clone are RRll 105 & RRIC 100. Tolerant to pink disease and abnormal leaf fall, but susceptible to powdery mildew.
7.	RRM 600	A high yielding clone evolved by the Rubber Research Institute of Malaysia and extensively grown in all the rubber growing countries. Parents are Tjir 1 and PB 86. Average annual yield per ha. in estates over 20 years is 1349kg. Latex unsuitable for concentration. Highly susceptible to diseases caused by *Phytophthora.*
8.	GTI	A primary clone developed in Indonesia and extensively planted in all rubber growing countries. Average annual yield obtained in estates for 19 years is 1,420kg per ha. Latex is white.
9.	RRII 5	A primary clone developed by the Rubber Research Institute of India. Selected from Malankara Estate, Thodupuzha.
10.	RRII203	A moderate yielding clone evolved by the Rubber Research Institute of India. Parents are PB 86 and Mil3/2. This clone exhibits high yield in Kanyakumari district.
11.	PB 28/59	A Malaysian clone with average annual commercial yield over 19 years is 1,477 kg per ha. The clone is highly prone to abnormal leaf fall, pink and powdery mildew diseases.
12.	PB 217	The parents of this Malaysian clone are PB 5/51and PB6/9. Average yield over the first 15 years from estates is 1508 kg/ha./year. Latex colour is light yellow.

13.	PB235	This Malaysian clone has PBS/51and PB 5178 as parents. Average annual yield obtained in estates over the first 15 years of tapping is 1501 kg/ha/yr. Latex colour is pale yellow.
14.	RRIM703	The parents of this clone are RRIM 600 and RRIM 500. Latex colour is light yellow.
15.	PR255	The parents are Tjir I and PR 107.
16.	PR261	The parents are PR 107 and Tjir 1 respectively. Mean yield in over ten years is 51.34g/tap/tree.
17.	PB 312	A Malaysian hybrid clone of parentage RRIM 600 x PB235. Shows high incidence of abnormal leaf fall and powdery.
18.	PB 314	A Malaysian hybrid clone of parentage RRIM 600xPB235.
19.	PB 255	A Malaysian hybrid clone of parentage PB 5/51xPB 32/36.
20.	PB311	The parents are PB 235 and RRIM 600 respectively. Average commercial yield over five years is 1450 kg/ha/yr.
21.	PB 86	Malaysian clone, yield 1130 kg per hector per year.
22.	PR 107	This clone developed in Indonesia. Average yield over 15 years under commercial planting is1,043 kg/ha/yr. Susceptible to *Phytophthora*. Tolerant to powdery mildew.
23.	TJIR-1	Indonesian clone, yield 930kg/ha/year and sensitive to Phytophthora, Oidium and pink disease.
24.	G.T.1	Indonesian clone, yield 1360kg/ha/year and tolerant to Phytophthora, Brown bast and pink disease.
25.	G.I.1	Malaysian clone, yield 1130kg/ha/year and sensitive to Brown blast.
26.	PRIM-600	Cross between TJIR-1 x P.B.86, Developed by Rubber Research Institute of Malaya (RRIM). Yield 1317kg/ha/year and sensitive to Phytophthora, and pink disease.
27.	RRIM.628	Cross between TJIR-1xRRIM.527, Developed by Rubber Research Institute of Malaya (RRIM). Yield 1051kg/ha/year and sensitive to Brown blast and poor yield during summer.

- For matured rubber trees under tapping apply NPK 10:10:10 grade mixtures at the rate of 900 g/tree (300 kg/ha) every year in two split doses. Add 10 kg commercial Magnesium sulphate for every 100 kg of the above mixture if there is magnesium deficiency.
- Growing of cover crops, incorporation of cover crops and weeding are important operations. *Pueraria phaseoloides, Calopagonium muconoides, Centrosema pubescens* and *Desmodium evalifolium* are common cover crops.
- Ethrel is recommended to increase latex yield of trees. It is applied at 5% a.i. concentration with a brush below the tapping cut to a width of 5 cm after light scraping of the outer bark. The first application may be done

after a drought period preferably after a few pre-monsoon showers and subsequent applications may be done in September and November. However, continuous application of Ethrel is not recommended for periods of more than 3 years at a stretch.

- Prolonged dripping of latex with the gradual decline in volume yield, pre coagulation of latex and partial or complete drying of tapping area (10-25 per cent). It will be control reduce tapping intensity or give a tapping rest for 3 to 12 months.

Tapping

A *Jeboug* knife is used, a knife with a V-shaped cutting edge which leaves a grooved channel for latex to flow. Present method is a single spiral 1/2 diameter of the tree. Cutting is repeated 1 to 3 days; 1/16" per cut, 7/8" per month, 10" per year. There are different systems on the length of the spiral and frequency of cut.

S/2 d/2	(half spiral, alternate day for 6 months and rested for 3 months)	100% Intensity
S /2 d/2 6m /9	(half spiral, alternate day for 6 months and rested for 3 months)	67 intensity
S /2d/3	(half spiral, third day)	67 intensity
S/2 d/31m/2	(half spiral, daily for one month and rested for next month)	100% intensity
S /1 d/4	Full spiral, fourth day	100% intensity
V /2 d/2 12m/16	Half circumstances and cut alternate day for 12 months and rested for the next 4 months	75% intensity

Yield:- India, the annual yield of rubber is 375 kg/ha from seedlings trees, whereas budded plants yield 800 - 1000 kg/ha.

Processing

There are three types of latex.

1. Liquid: Highest quality
2. Cup rubber: The cut-lump that sticks to the plastic cup
3. Scrap rubber: Strip of rubber from the cut surface.

    - Coagulation 10–25% ammonia can be added as an anticoagulant La tex bulked and strained, rubber content 30–35%. Latex diluted with water, poured into coagulation tanks, with acetic or formic acid (1 part 4% formic acid required for 100 parts 12% latex more if ammonia is added).
    - The "curd" is separated from the serum, and squeezed in rollers to 2-1/2 mm thickness, dried in smoke house, bailed for export. Crepe rubber is compressed granular rubber from sheet rubber that passes through rollers. Crumbing is the operation that cuts up crepe and compresses into blocks.

Pests

Pests	Control
Scale insect	When severe infestation is noticed, spray Organophosphorus insecticides like malathion 50 EC 2 ml/lit.
Mealy bug	Spray fish oil rosin-soap 25 g/lit. Release Australian lady bird beetle, *Cryptolaemus montrouzieri* @ 10/tree.
Termite (White ant)	Drench the soil at the base of affected plants with Chlorpyriphos 20 EC 2 ml/litre.
Cockchafer grub	Drench soil at the base of plants in the affected area with the solution of Chlorpyriphos 20 EC 2 ml/litre.
Mites	Spray Sulphur 50 WP at 2 g/lit or spray Dicofol 18.5 EC 2.5 ml/lit.

Diseases

Diseases	Control
Abnormal leaf fall	Prophylactic spraying on the foliage prior to the onset of South-West monsoon with, Bordeaux mixture 1% at 4000 - 5000 lit/hausing high volume sprayers.
Secondary leaf fall	Prophylactic spraying on the foliage prior to the onset of South-West monsoon with, Bordeaux mixture 1% at 4000 - 5000 lit/ha using high volume sprayers.
Powdery mildew	Dusting during the defoliation period commencing from the bud break

Diseases	Control
	in about 10% of the trees, giving 3 to 5 rounds at weekly to fortnightly intervals before 10.00 a.m. using 11 to 14 kg 325 mesh fine Sulphur dust per round per ha. Sulphur dust can be mixed with talc in the proportion of 7:3. Wettable sulphur (1 kg in 4000 lit of water) is also effective in nurseries and for young plants as a spray.
Bird's eye spot	Repeated sprayings with Bordeaux mixture 1% or Mancozeb or Copper oxychloride 0.2%. Provide shade in nursery. Give balanced manuring to increase tree vigour.
Leaf spot	Spray 1% Bordeaux mixture or 0.2% Mancozeb, or 0.1% Carbendazim at fortnightly intervals.
Pink disease	Frequent tree to tree inspection should be done during July – September period for detecting the infected trees and application of Bordeaux paste in the early stages upto 30 cm above and below the affected region.
Patch canker or Bark	The affected region may be scraped to remove all the rooting bark cankers and the coagulated rubber and the wound washed well with Dithane M 45 @ 0.75 %. When the fungicide dries up apply wound dressing compound.
Dry rot, Stump rot	Clean up affected areas, by washing with Carbendazim (0.1%) Collar rot or Charcoal solution. Scrape out the fructifications. Affected bark and wood **rot** show black lines. Wash the wound again with fungicide solution. When it dries up apply a wound dressing compound. Avoid accumulation of rubber at the base of the trees.
Brown root disease	Open up the root system. Completely killed and dried roots may be traced and pruned. Partially affected and healthy roots washed with fungicide Carbendazim (0.1%) solution. When the fungicide dries up, a thin coating with a wound dressing compound may be given. Refill the soil and drench the base with fungicide solution.

Chapter 8

Coffee

Two species mostly grown worldwide

Arabica Coffee	**Robusta**
Coffea arabica	*Coffea canephora*
It is a Bush.	It is bush-like tree.
Chromosome No. = 2n=44 (4x), Tetraploid	**Chromosome No.** = 2n=22 (Diploid)
It is a self-fertile plan.	It is a self incompatible plant.
Origin:- Ethiopia.	**Origin**:- Rain forest of Congo basin
Best quality.	Less flavor and acid.
Grown at >1600m height (High elevation).	Grown at <750m height (Low elevation).
Leaves are dark green.	Leaves are pale green.
It is require 15-24°C temperature and 1300 mm rainfall.	It is require 24-30°C and 1550 mm rainfall.
Inflorescence:- Cyme	**Inflorescence:-** Cyme
Flowers are scaly, small bracts, per axil 4-5 inflorescence of 1-4 flowers per inflorescence.	Flowers are leafy, expanded with 5-6 flowers per inflorescence.
10-20 berries per node. Berries are oblong to round in shape.	40-60 berries per node. Berries are small.
Fruit development in 8-9 months.	Fruit development in 10-11 months.
Blossoming after 9-10 days of the receipt of . blossom showers	Blossoming after 7 days of the receipt of blossom showers.
Flowers only on previous-season's growth.	Flowers on current season's growth.
Self-pollinated.	Cross-pollinated (wind and insect pollinated)
Share 70% of world production.	Share 30% of world production.
Main growing areas are Latin America, East Africa.	Main growing areas are Asia and Africa.
Caffeine content is 0.6 – 1.5%.	Caffeine content is 2.0 – 2.7%.
Susceptible to rust and berry borer.	Resistant to rust berry borer.

Type of Inflorescence: Racemes

- Coffee drinking first became popular in Yemen in the 15th century.
- Coffee derives its name from Arabic language.

- Caffelite, a type of plastic, can be made from coffee beans.
- A small, evergreen tree, reaching 30 ft in the wild, but generally maintained 6-8 ft in cultivation.
- Flowers are white, fragrant, ½ – 13 across, with 5 linear petals fused into a slender tube at their base.
- Fruit an epigynous berry.
- Coffee fruit are often called "cherries" and erroneously considered to be drupes.
- Fruit are ½ – 3/43 long, ovoid, and borne axillary in clusters on growth from the previous season (Arabica) or current season (Robusta).
- Single-seeded fruit occur, and are called pea berries.
- The ovaries develop into an oval fruit containing two seeds, and needs 6-11 months to ripen.
- Ripe coffee cherries have a red or yellow, sweet type of flesh; the actual coffee beans are contained within them, in a membranous pellicle and thin, hard endocarp.
- In comparison with Arabica, 30 % higher yields are gained from Robusta, although the price is around 30% lower.
- *Coffea maragohipe*, which has extremely large coffee beans.
- Coffee was introduced into India during 1600AD by a Muslim Pilgrim, Baba Budan on the hill near Chikmangulur.
- Orthogeotropic commonly called suckers which grow vertically.
- Plagiogeotropic branches, commonly called primaries which are placed at 45-90° in relation to the main axis.
- Primary branches give rise to secondary branches which split to tertiary branches and which also branch to form quaternary branches.

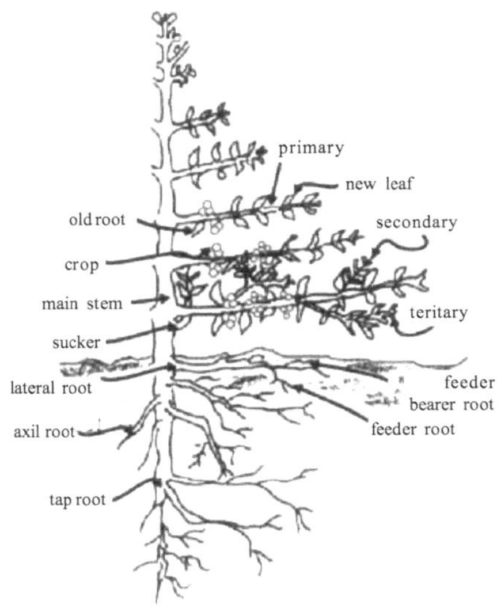

Morphology of coffee plant

- Suckers which originate after part of the main stem is excised are axillary. Sucker regeneration can be induced by partial ringing of the bark tissue just above the point of stumping. Sucker development was found to be under the influence of apical dominance.
- Commercial cultivation started in South India in 1820 by British entrepreneur.
- Coffee is used almost exclusively in the drinks industry, and is offered to consumers as roasted beans, ground, and also as instant coffee.
- Coffee is short day plant.
- Coffee is a half-shade plant, that can only utilize around 1% of the sunlight (ideal is around 1500 hours per year) photo synthetically.
- Coffee is one of the major plantation crops mainly grown in Southern States [Karnataka (56%), Kerala (25%) and Tamil Nadu (9%)] and to a small extent in Northeastern States of India.
- Coffee cultivation is confined mostly to hilly tracts of Western and Eastern Ghats in India.
- Arabica is generally seed propagated.
- A gentle sloping land without big shade trees is preferred for raising nurseries.
- For sowing the coffee seeds, germination beds of 1 metre width and of convenient length raised to a height of 15 cm from the ground level should be prepared.
- The seeds are treated with Agrosan or any Organomercurial compound to prevent fungal infection.
- Root cuttings are used for Robusta propagation.
- Low multiplication rate is the main limitation for a large-scale field adoption of root cutting method.
- Wedge-cleft grafting at seedling stage (2 months old) with Arabica scion and Robusta rootstock is recommended for planting in nematode-infested areas as Robusta is relatively tolerant to nematodes.
- **Planting time** is June – December.
- **Planting distance:** (Tall varieties) 6-7 x 6-7 feet

  (Semi-dwarf varieties) 5 x 5 feet

- **Robusta:-** 8-10 x 8-10 feet.

  Pit size:- 45 x 45 x 45 cm.
- After the receipt of one or two summer showers, coffee plants start producing new shoots. This is the right time for pruning.
- Red and lateritic soil suitable for coffee production.

Maturity Index: Individual fruit turn from green to completely bright red and glossy over a period of 1-2 weeks when mature. The seeds are easily squeezed from mature fruit by moderate finger pressure.

- Coffee fruits have to be picked as and when they become ripe; this can be understood by gently squeezing the fruit with fingers. On gentle squeeze, the bean inside the fruit pops out easily.
- Under ripe or over ripe fruits should not be mixed or used for the preparation of parchment coffee as they can spoil the overall cup quality.
- The under ripe coffee fruits result in the production of immature beans and the over ripe ones cause foxy coffee resulting in brown beans with fruity cup.
- The under and over ripe fruits can be converted into cherry coffee.
- The coffee crop is picked by hand, a labour-intensive and difficult process.
- While harvesting the coffee, picking mats should be used, this helps in easy collection of fallen fruits and prevents the mould formation.
- Prevention of mould formation avoids the production of Ochra Toxin-A in coffee beans.
- Use of picking mats also reduces the Coffee Berry Borer infection.

Preparation of Parchment coffee:- Coffee processing has different steps, that are under:-

Pulping

Preparation of such coffee requires pulping equipments and adequate supply of clean water. The quality of parchment coffee is always superior to that of cherry coffee (whole fruit dried under sun). Cherry coffees always give fruity taste as the beans are in contact with the mucilage for a longer time during the course of drying period. Therefore, wet processing of coffee is essential to get superior quality coffee.

The pulper, pulper tanks, siphon tanks, rakers, washer, fermentation tanks should be thoroughly cleaned before the pulping operation is undertaken. The premises of the pulper unit as well as the drying yards should be kept in hygienic way. Only clean water shall be used for processing coffee fruits and all the extraneous matters such as leaves, twigs and stones should be thoroughly removed from the fruits before pulping.

As far as possible, harvested fruits have to be pulped on the same day of the harvest. If, it is not possible to process all the fruits on the same day, such fruits can be stored under clean water and pulped on the second day. The fruits stored for more than 10 hours in the open heap without pulping will result in fruity cup. The fruit skins separated by pulping should be taken away from the pulping site as quickly as possible so that microbial decomposition of the skin does not affect the quality of parchment coffee.

Demucilaging and Fermentation

The sticky mucilage on the pulped beans can be removed by 4 different methods. However, the commonly followed/adopted methods are two namely: (1) Natural Fermentation and (ii) Aqua machine wash by friction.

Superior quality coffee can be obtained by natural fermentation method. Fermentation is a critical stage in processing and has a deep influence on the quality of coffee. When pulped coffee is stored in fermentation tanks along with the mucilage, natural hydrolysis of pectins present in coffee mucilage occurs due to the enzymes naturally present in coffee mucilage. In Arabica coffee, natural fermentation will be complete in about 24 to 36 hours. However, this may take longer time also when the ambient temperature is too low. The fermentation process should be controlled so that over fermentation is avoided. Over fermentation will result in foxy beans and make the coffee cup sourish. The under fermented coffee results in imparting mustiness in the cup. The correctness of fermentation can be checked by rubbing the fermenting beans with hand. The mucilage will come off easily and the parchment does not stick to the hand after wash. The beans feel rough and gritty when squeezed by hand and a feeling similar to squeezing pebbles is felt.

Alkali Treatment

Removal of mucilage by alkali treatment takes one hour for Arabica and one and a half to two hours for Robusta. The beans obtained after pulping are drained of excess water and spred out in vats uniformaly. A 10% solution of caustic soda (sodium hydroxide) is evenly applied over the beans and stirred using wooden ladles and trampled by feet for half-an-houre.

Washing

After the correct fermentation, the beans are cleanly washed 3 to 4 times in clean water or in a washing machine. The quality of the coffee can be further enhanced, if the washed beans are soaked under water for about 4 to 6 hours.

Drying

The parchment coffee prepared in such method initially contains around 50 – 55% water content in them. The superficial water content should be removed by spreading the parchment over wire mesh trays and then dried on clean cement or clay lined drying yards till the moisture content reaches a value of 11.0%. Generally parchment takes about 7-10 days for drying when spread at a thickness of 4-7 cm. If the parchment is spread too thin, rapid drying of coffee at this stage causes the parchment skin to split which lead to cracking of parchment, shrunken and boat shaped beans. Stirring and turning of parchment coffee at least once an hour is essential to facilitate uniform drying. The parchment should be heaped during the evening time and covered with polyethylene/tarpaulin sheets/gunny bags. Drying mass of parchment coffee should never be exposed to mist. Parchment coffee should not be exposed to the hottest sun between noon and early part of the afternoon especially on the 3rd and 4th day of drying. Exposing the parchment to the hottest sun during the 3rd and 4th day will result in wrinkled beans. Once the coffee is dried to a recommended moisture level of 11.0%, the coffee has to be bagged in clean and dry gunny bags. Coffee of different lots should be bagged separately. New gunny bags should be always turned inside out well aerated before use as otherwise the coffee will absorb natural oils and off odours from the bag and give rise to an "acrid" (bitter/pungent/harsh) cup. If the gunny bags are procured from curing works, the bags should be properly fumigated or the curing works should be asked to fumigate and supply to avoid the spread of Coffee berry Borer infestation.

Storage

Coffee should be stored in well ventilated and dry store rooms. The coffee should be stored on raised wooden platforms to ensure proper circulation of air underneath the bags.

Preparation of Cherry coffee

To prepare cherry coffee, fruits should be picked as and when they are ripe. Greens and under ripe fruits should be sorted out and dried separately. All the extraneous matters like twigs and leaves should be separated from the fruits and dried. The fruits should be spread uniformly in the drying yards at a thickness of 7–8 cm. Clean cement or clay lined drying yards should be used for drying the cherry coffee. Use of cow dung smeared or mud drying yards should be avoided. Drying coffee in such yards causes mould formation in cherry coffee and quality of such coffees is very inferior. Coffee should be stirred and ridged at least once every hour to facilitate uniform drying. In the evening time cherry coffee should be heaped and covered with polythene or tarpaulin sheets to avoid contact with the mist. The coffee will be fully dry in about 10 -12 days under the bright weather conditions. Cherry coffee should be dried to a moisture level of 12 to 12.5%. The cherry is dry when handful of the drying cherry produces rattling sound when shaken and a sample for lit records the same weight on two consecutive days. At the end of the harvest, the remaining fruits on the plants should be completely stripped and fallen fruits should be gleaned cleanly. If this method of complete stripping and gleaning is not followed, there is a danger of Coffee Berry borer infestation during the coming years. Stripped and gleaned coffee should be dried separately and should not be mixed with other coffee.

Curing: Curing is done in curing house established by licensed curers. It refer to processes like bulling (removal of fruit wall and parchment), grading and sorting.

Blending: Coffee beans after curing are roasted and ground to get coffee powder. Coffee from different region differ in quality and hence arabica is known for judicious blending aroma and robusta for strength. *Chicorium intybus* is used as an additive. It improves colour, odour and taste of coffee. This combination is known as 'French coffee'.

Speciality Coffees

Flavoured coffee: Coffee is flavoured with flavour of cardamom, cinnamon, lemon and saffron.

Decaffeinated coffee: Green beans are extracted with solvent like ethyl acetate to reduce caffeine content to less than 0.1%.

Instant coffee (Soluble coffee): The concentrates of coffee are extracted in hot water at 90-150°C, for 3 times, in steel vessels. The coffee extract is spray dried/freeze dried.

Arabica Coffee Varieties

Varieties	Parentage	Special characters
Selection 1 (S288)	Tetraploid hybrid from S26 (natural cross – *C. liberica x C. arabica*).	Resistance to leaf rust race 1 & 2, high yielding. Wider adaptability.
Selection 3 (S795)	Cross between S.288 x Kent.	Resistance to leaf rust race 1 and 2, 700 -1200kg/ha Bold fruits 75% A grade.
Selection 5	Devamachy x S -881.	Small, oblong fruits 900 – 1100kg/ha.
Selection 6	S-274 (Robusta) x Kents.	High A grade beans, 900 – 1000kg/ha.
Selection 7	San Ramon –a short inter-nodes Arabica.	Dwarf in stature, segregates to tall by 30%.
Selection 8	Pure line selection of *Hibrido –de -Timor.*	*Highest resistance to leaf rust, rooping branches.*
Selection 9	Sln 8x Tafarikela.	Drought hardy, suitable to different coffee zones.
Selection 10	Double cross hybrid, Cattura x(S795 XHDT).	Resistant to leaf rust.
Selection 11	Progency of *C.liberica* X *C.eugenoides.*	Field resistance to rust and drought hardiness.
Cauvery	Catimor x Hibrido de Timor.	Plants are dwarf, suitable for high density plantingYield – 3000kg/haMore a grade beans with superior cup quality.
Selection 12	Cross Caturra x HDT, followed by selfing.	Precocious, suitable for close planting, resistant to leaf rust.
Sln 274		
Sln 270		
Sln 3		

Robusta Coffee Varieties

Varieties	Parentage	Special characters
Sel-1R (S-274)	This is a single plant progeny giving 1400-2500 kg clean coffee/ha.	Resistance to leaf rust race 1 & 2, high yielding. Wider adaptability.
Sel-2R (S-270)	It is selection from Robusta.	Resistance to leaf rust race 1 and 2, 700 -1200kg/ha Bold fruits 75% A grade.
Sel-3R	It is an inter-specific between *C.congensis* X *C. canephora*	Small,oblong fruits 900–1100 kg/ha

Pests

- White stem borer: *Xylotrechus quadripes.*
- Cockchafers or white grubs: *Holotrichia* spp.
- Coffee berry borer: *Hypothenemus hampei.*
- Hairy caterpillars: *Eupterote* spp.
- Shot hole borer*: Xylosandrus compactus.*
- Coffee bean beetle: *Araecerus fasciculatus.*
- Mealy bugs: *Planococcus* spp.
- Snail: *Ariophanta solata.*
- Green scale: *Coccus viridis.*
- Red borer*: Zeuzera coffeae.*
- Brown scale: *Saissetia coffeae.*

Disease

- **Leaf rust** (*Hemilcia vastatrix*)**:** This is an important disease causing economic losses particularly in arabica coffee on the lower surface of infected leaves.
- **Black Rot or Koleroga Rot** (*Koleroga noxia*)**:** It is affected both type of coffee during rainy season and high relative humidity times.
- **Anthracanose** (*Colletotrichum gloeosporioides*)
- **Berry Blotch** (*Cercospora coffeicola*)
- **Root Root** (*Rosellina arcuata* and *R. bundoes*)
- **Fusarium root** (*Fusarium oxysporum*)
- **Brown blight**
- **Stalk rot of berries and leaves**
- **Red cliffe disease.**

Physiological Disorder

- **Kondli or Stem wasting:** It is due to copper toxicity in young plant.

Chapter 9

Palmyrah Palm

Botanical Name: *Borassus fabellifier*

- *Borassus flabellifer* - Asian Palmyra palm/Lontar palm/Doub palm/Sea Coconut (Southern Asia and Southeast Asia)
- *Borassus aethiopum* - African Palmyra Palm (Tropical Africa)
- *Borassus akeassii* - Ake Assi's Palmyra Palm (West Africa)
- *Borassus heineanus* - New Guinea Palmyra Palm (New Guinea)
- *Borassus madagascariensis* - Madagascar Palmyra Palm (Madagascar)
- *Borassus sambiranensis* - Sambirano Palmyra Palm (Madagascar)

Origin: Tropical Africa, Asia and New Guinea

Family: Palmae/ Arecaceae

Chromosome No.: 2n = 2x = 36

Type of Fruit: Drup.

Type of Flower: Spadix

- The palmyrah palm has long been one of the most important trees of Cambodia and India.
- Palmyrah palm was mentioned in the early historical writings of Manuneedhi Sastra.
- Chinese traveler Magesthanese recorded the tapping of Padaneer and making of sugar candy which he took back his country.
- Palmyrah palm is also known as 'Kalpaga tharu' because it's all part are usable. It is also known as 'Tree of life' because it useful in food and shelter.

- The word 'Borasus' was derived from a Greek word and it means the leathery covering of the fruit and 'Flabellifer' means fan-bearer.
- In Cambodia, the tree is a national flora symbol/emblem that is seen growing around Angkor Wat.
- Palmyrah Palm is also known as brab tree, toddy palm, Tala Palm, tar-ka-jhar and Wealth of India.
- The Tamil Nadu government also declares as a State tree in 1978. Highly respected in Tamil culture, it is called "Karpaha Veruksham" ("celestial tree") because all its parts have a use.
- Palmyra palms is evergreen tree.
- Palmyrah palm mostly grown in 44° South latitude to 45° North latitude.
- In India mostly grown in Southern State Tamil Nadu, Andhra Pradesh, Kerala, Orissa, West Bengal etc.
- Tamil Nadu is leading state in area and production Palmyrah palm.
- Palmyrah palm belong to sub-family Borassoideae.
- Palmyrah palm dioecious character (male and female flower at separate plant).
- These are tall palms (grows up to 30m), leaves are long, fan-shaped, 2 to 3 m in length, flowers are small and normally found in densely clustered spikes and fruits are large, brown and round shaped
- A well mature palm on an average bear 50 fruit in bunches. Fruit is normally yellow to light orange with black bottom. Fruits are normally seeded and mesocarp is pulpy and fibrous.
- Flowering appears generally in March and fruit bear in May to August.
- The Palmyrah palm is generally cultivated in tropical region but it is also grown in some sub-tropical.

Soil: Arid, deep sandy & loamy soils are suitable. Dry areas with low rainfall are also ideal.

Propagation: by seed.

- Seed germination of 63.25 % are achieved 6 weeks after sowing.
- Palmyrah palm's seed is poly-embryonic nature generally 2 seedling emerge from one seed.

Nursery transplanting: About 1 year old seedling is lifted from the nursery and containerized in polythene bags.

Spacing: 3m x 3m (1110 palms/ha)

Pruning: Pruning of 30 per cent leaves @ 10 per cent in phased manner.

Growth: Palmyrah is a slow grower. First frond appears in about 5 months. First fan shaped tree leaves appears only in the 2nd year. When it attains the height of 12–18 m, comes to flowering (13–15 yrs) for *Padaneer* (Sweet sap) purpose. Average of 100 – 200 lit obtained for a period of 4 months from Feb – May.

Harvesting: Fruits are harvested by hand. Climb the tree and cut clusters of fruits with a hatchet or big knife. It may take 20 years until the first harvest, but the tree can get more than a hundred years old.

Yield: 150 litres of padaneer / tree / year

- 24 Kg jaggery/tree/year.
- Jaggery recovery/litre of padaneer: 180 – 250 g of jaggery.
- The tree also yields many types of food. The young plants are cooked as a vegetable or roasted and pounded to make meal.
- The fruits are eaten roasted or raw, and the young, jelly like seeds are also eaten.
- The jelly like seed is a famous and delicious summer food; the British called this tender fruit Ice-Apple.
- The season of Palmyra fruit is May.
- A sugary sap, called Neera, can be obtained from the young inflorescence, either male or female ones. It is used as a stimulant and antiphlegatic. It is also useful in inflammatory infections and dropsy.
- Toddy is called "*Kallu*" in Telugu.
- The fresh sap contains 12% saccharose.
- The Neera is fermented to make a beverage called Toddy. The alcohol content in fully fermented toddy is estimated to be about 5%.
- Palmyrah arrack obtains from toddy by pot still distillation and patent still distillation.
- The roots can be dried to form Odiyal, a hard chewable snack.

- The tree sap is taken as a laxative, and medicinal values have been ascribed to other parts of the plant.
- The collected sugar sap is concentrated to make a crude sugar called Jaggery (It is mainly done in Java Island, Indonesia). It is called "Gula Jawa" (Javanese sugar) in Indonesia and is widely used in the Javanese cuisine.
- Palm shoot is cut and the juice is traditionally collected in hanging earthen pots. The juice so collected before morning is refreshing and light drink called "*Thaati Kallu*" in Telugu, Neera in Marathi and "*Pathaneer*" in Tamil is extremely cool in sensation, and has a sugary sweet taste. The juice collected in evening or after fermentation becomes sour is called Tadi in Marathi.
- Tadi is consumed by coastal Maharashtra mostly by villagers as a raw alcoholic beverage.
- Palm sugar also prepared from sap and it is substitute of cane sugar.
- Palmyrah treacle is a substitute of sugar; it is used as sweetener for curd and bread. It is golden colour and viscous nature.
- Palm cola is an aerated soft drink containing 11% sugar.
- Tender Palmyrah fruit is called Nangu.
- Palmyrah fiber is extracted from bifurcated portion at the base of the leaf stalk joining in canoex of the palm.
- Neera is the fibrous portion extracted from the stalk of fronds.

Tapping: Extraction of sap (Neera/ Padaneer) from inflorescence is called tapping. According to sex of the palm & age of the inflorescence, different kinds of tapings are available.

Aripanai: In male palm sheath covering the young inflorescence is removed and dried for 3 weeks. End is cut every time & pot is tied (1-1½ months).

Vallupanai: One month old spikes are selected. Each male spike bearing sessile flowers is pretreated by pressing and stroking & 3 - 6 such spikes are brought together, wrapped with leaves and fitted to a pot.

Thattupalai: It has to be done in female palm to soften the tissue by hitting the inflorescence main axis with iron rod.

Kaivetty: Employed when the inflorescence is about 2 – 3 months old. Female palms are tapped for a longer period. i.e., April- Dec. and male for Dec – Feb.

Sap is collected twice a day. Each time at the end of the collection of sap, a new cut surface is made by thin slicing. Tapped sap is called Neera or Padaneer.

Variety

SVPR 1 : It is a selection from Srivilliputhur local. It is semi tall in height and tapping duration is upto 95 days. It is suitable for growing in entire Tamil Nadu. It yields 298 liters of Padaneer/tree/year. The Jaggery recovery is higher (144 g Jaggery/lit of Padaneer).

Chapter 10

Oil Palm

Botanical Name: *Elaeis guineensis* Jacq

Family: Arecaceae

Chromosome No.: 2n= 2x=-32

Origin: Tropical Africa.

Type of Fruit: Drupe

Type of Inflorescence: Spadix

Useable Part:-Fruit.

- Oil palm is recognized as the highest and cheapest edible oil-yielding crop.
- Oil palm yield of edible oil is 5.5-6 tonnes per hector.
- Palm oil is widely used as cooking oil, as an ingredient in many processed foods and as a substitute for butter.
- Palm oil can be used to produce biodiesel.
- Palm oil is also used as a binding agent in cosmetics.

Classification of Oil Palm based on fruit structure

1. **Dura:-** Shell usually 2-8 mm thick, low to medium mesocarp content (35-55%) kernels large, no fiber ring.
2. **Tenera:-** Shell 0.5-4 mm thick, medium to high mesocarp content (60-90%) fiber ring darker in colour and encircles the endocarp. Higher sex ratio and larger number of bunches than dura.
3. **Pisifera:-** Shell less with small pea like kernels in fertile fruits. It is of little commercial value, but is important commercial palms.

Classification of Oil Palm based on the colour of exocarp

1. **Nigrescens:-** Unripe fruit deep violet to black at apex and ivory coloured towered base. This is the common type. Two forms are recognized on ripening; Rubro nigrescens:-Ripe fruits deep reddish-orange, it has the highest content of carotenoids and carotene. Rutilo nigrescens:- It is ripe fruits paler-orange with black cap on upper half.
2. **Virescens:-** Unripe fruits green, ripening to light reddish-orange with small greenish tip. Anthocyanins little or absent.
3. **Albescens:-** Fruit lack reddish colour at maturity as it contains little or no carotene. It ripens to pale-yello or ivory with a blackish or green cap on upper half.

- Oil palm can reach 60-80 ft in height in nature, but is rarely more than 20 or 30 ft in cultivation.
- Oil palms are monoecious, producing male and female inflorescences in leaf axils.
- The inflorescence of both sexes is a compound spadix with 100-200 branches, initially enclosed in a spathe or bract that splits 2 weeks prior to anthesis.
- Oil palms are primarily pollinated by weevil.
- The oil palm tree is a tropical plant which commonly grows in warm climates at altitudes of less than 1,600 feet above sea level.
- Oil palm is a humid crop. Requires evenly distributed rainfall of 150mm/month or 2500-4000mm/annum.
- Crop comes up well between 29-33°C max. and 22-24°C min. temperatures and with bright sunlight for at least 5 hrs. per day.
- Humidity of more than 80% is required to come up well.
- Best-suited soils are moist, well-drained, deep, loamy alluvial soils, rich in rganic matter with good water permeability. Avoid highly alkaline, highly saline, waterlogged and coastal sandy soils.
- Palms will grow in a variety of deep tropical soils and will tolerate a pH between 4.0 and 8.0.
- Oil palms have a requirement for continuous moisture throughout the year.
- The trees can tolerate 2 to 3 months of dry conditions but it can affect yields.

- Oil palm is propagated from seed.
- 12 -14 months old healthy seedlings with 1-1.3m height and 13 functional leaves are recommended for planting.
- Planting distance should be 9m x 9m x 9m (triangular planting).
- Planting should be done in pit size of 60 cm x 60 cm x 60 cm.
- For grown up yielding palms of 3 years age and above, a minimum of 150 to 200 liters of water per day is required.
- Oil palms are frequently planted along with a leguminous cover crop which helps provide nitrogen to the palms while also suppressing weeds.
- When bunches are harvested, it is usual for fronds to be pruned.
- The number of fronds per palm is maintained at between 35 and 40.
- Glyphosate (750ml/ha/year or 17.5 ml/basin) is recommended for effective weed control.
- Oil palm comes to flowering 14-18 months after planting. It produces both male and female flowers separately on the same palm.
- Ablation is the removal of male and female flowers produced in the early stages of plantation. This enables the plant to gain adequate stem girth, vigour and develop adequate root system. Flowering starts from 14th to 18th month after planting. Start ablation immediately after the appearance of inflorescences on the palms. They can be removed easily by hand pulling or using the tool developed at DOPR. Ablation can be extended up to 2.5 to 3 years depending upon the plant growth and vigour.
- Oil palm is a highly cross-pollinated crop. Wind and insects assist pollination, but wind pollination is not adequate. Effective pollinating insects like *Elaeidobius kamerunicus* helps in good pollination and fruit set.
- Harvesting of oil palms generally begins 2.5 years after the trees are planted in the field.
- While harvesting a stalk length of 5 cm alone should be left. Harvesting should be done at 10-12 days interval. During rainy season, harvesting should be done at closer interval of 6-7 days as ripening is hastened after hot summer.
- Palm oil cake contains 18-19% protein.
- Palm wine is obtained by tapping the male inflorescence.

Diseases

Diseases	Causal organism	Symptoms	Control
Spear Rot	Mycoplasmal		
Pestalotiopsis leaf spot	*Pestalotiopsis* spp. (Fungi)	Yellowing start form the tip and speared on whole plant.	In early stage, the affected leaves should be removed and burnet.
Bud rot or little leaf Diseases		Rotting with brown discolouration start from the lower surface of unopened spear leaf and spreads from one spear to another.	Spraying of 1% Bordeaux mixture.
Ganoderma bud rot	*Ganoderma* spp.		There are currently no fungicides recommended for protecting palms from Ganoderma bud rot.
Trunk rot		Young leaves fail to open because a number of fully elongated but unopened spears are seen in the centre of the crown.This diseases associated with drought in potassium deficient soils.	
Fruit Rot	*Marasmium palmivora*		
Anthracnose		It is occure in the nursery.	
Bacterial bud rot	*Erwinia* spp.		Plant oil palm varieties with resistance to the bacteria; palm buds can be protected using copper-based fungicides
Oil palm wilt	*Fusarium oxysporum*		International quarantine procedures have limited the spread of the disease between major palm oil producing countries

Insect-Pest

Rhinoceros beetle: *Oryctes rhinoceros*

Mealybugs (Pineapple mealybug): *Dysmicoccus brevipes*

Bag Worms

Slug Caterpillars

Physiological Disorder

Crown diseases: The folded leaflets being to rot. It is due to deficiency of potassium and magnesium.

Oil Palm Processing

Processing of palm oil can be divided into 3 major stages Extraction, Refining and Fractionation.

1. Extraction: The serves the dual objective of inactivation enzyme lipase and loosening of fruits from the bunch. It also softens the cell wall and coagulates proteins, facilitating oil extraction in subsequent stage. Horizontal type sterilization is used to minimize bruising of fruits. The FFB is loaded in a perforated cage and placed in the sterilizer. Sterilization for 60 minutes at 45°C is optimum. Sterilization of the bunch are important, since the complete natural loosening of the fruits takes a week or more, resulting in increasing in free fatty acid (FFA) lowering the oil quality. Stem sterilization of fruit bunches at 3 Kg/cm^{2}pressure for 30-60 minutes is done to arrest the development of FFA and also to soften and loosen the fruits for easy stripping and pounding.

Stripping: Sterilization fruits are separated from the bunch by passing through a rotary drum stripper (20 rpm) with baffles and perforations.

Digestion: The purpose of digestion is to detach the fruit pulp and get the cell walls broken, facilitating release of oil with the help of thermal and mechanical energy. The digester is a vertical jacketed cylindrical vessel fitted with a centrally mounted agitator having specially designed blades which rotate at slow speed (25 rpm). The loose fruits are charged into the digestor from the top. Live steam is injected into the vessel to maintain the temperature at 95°C. The digested mash with semi-solid consistency is discharge chute at the bottom.

Pressing: The digested mash consisting of oil, water, seed, fiber and other suspended matter is charged into a perforated cage and subjected to pressing. The hydraulic press if fitted with a plunger matching with the cage diameter. The pressing is done while the mas is hot (80-90°C) at 750 psi. the hot oil-water mixture with suspended solids is expelled through perforation leaving the solid press cake in the cage.

Clarification: the oil water mixture is filtered to remove the fibrous debris. It is collected in the clarification tank. Clarifer is a vertical cylindrical vessel fitted with steam coil. The oil- water mixture is diluted with hot water (1:2) and heated at 95° C. The oil being lighter rises to the top and is decanted continuously into a collection tank. The watery sludge at the bottom is discharged as waste.

Purification and Storage: The crude palm oil from the clarifier is passed through a high-speed centrifuge at 80°C to remove the trace of solid impurities and water. The final trace of moisture is removed by vacuum-drying and pure raw palm oil is storage in tank.

2. Refining: Refining of crude palm oil (CPO) is carried out for removing impurities, free fatty acids, colour and odour. This is achieved through 2 routes, chemical refining and physical refining.

Degumming: It is the process of precipitating impurities such as phosphatides, protein fragments and mucilaginous matter. The degummed oil is treated with calculated amount of caustic soda to remove free fatty acid in the form of soap. The soap stock is separated by centrifugation and neutralize oil is washed free of alkali and soap using hot water at 90°C followed by vacuum-drying.

Bleaching: It is done to obtain palm oil with lighter colour and to remove traces of soap. Depending upon the quality of CPO 1 to 3% bleaching earth at temperature ranging from 100-150°C under vacuum for 15-30 minutes is employed. Oil is pumped through a filter press to separate the spent earth.

Deodorization and polishing: this is achieved by passing superheated steam at 230-240°C under vacuum for 2 hours. When fully deodorized, the oil is cooled to 55°C and pumped through a polishing filtrate to give the oil its final sparkle.

Physical Refining: crude palm oil is first mixed with 0.1% phosphoric acid followed by bleaching with 1-2% earth under vacuum at 100-150°C. After suitable content time, the spent earth is removed by filtration. This degummed and bleached oil is removed by filtration. This degummed and bleached oil is deaerated and stem-stripped under vacuum at 240-270°C. Finally, the oil is cooled and passed through a polishing filter. Physical method yield better quality oil and avoids pollution. The end- product of both refining methods is the same.

3. Fractionation: The refined palm oil is separated into liquid fraction or palmolein and soiled fraction by different processes such as crystallization detergent fractionatio or solvent fractionation.

Chapter 11

Betelvine

Botanical Name : *Piper betel*

Family : Piperaceae

Origin : Malaysia.

Chromosome No. : 2n=2x=26

Useable Part : Leaf

- Betelvine leaves have a strong pungent aromatic flovour and are widely used as masticatory.
- Mature leaves are used for chewing with smeared hydrated lime plus catachu, arecanut, clove, cardamom, clove, funnel etc.
- Betal chewing is considered as a good and cheap source of dietary calcium.
- Betal oil has several medicinal used it used for brain leaves carmination, etc. Leaves has religious importance.
- Moisture 85.4%, protein 3.1%, fat 0.8%, carbohydrate 6.1%, fibre 2.3%, minerals 2.3%, reducing sugars 0.38%, to 1.46% and all vitamins and Iodine also.
- The aromatic value is determined by the nature of oil content in leaf. Oil content varies from 0.7 to 2.6%, and the oil consist of phenols and terpens.
- Betalvine thrives best under humid tropical condition with cool shade. It flourishes in areas with a rainfall of 225 to 475 cm.
- The betel cultivation in clay loam which is friable, rich in organic and which has good drinage.

Important betelvine varieties

- **Andhra Pradesh:** Karapaku, Chennor, Tellaku, Bangla and Kalli Patti.
- **Assam:** Assam Patti, Awani pan, Bangla and Khasi Pan.
- **Bihar:** Desi Pan, Calcutta, Paton, Maghai and Bangla.
- **Karnataka:** Kariyale, Mysoreale and Ambadiale.
- **Kerala:** Nadan, Kalkodi and Puthukodi.
- **Madhya Pradesh :**Desi Bangla, Calcutta and Deswari.
- **Maharashtra :**Kallipatti, Kapoori and Bangla (Ramtek).
- **Orissa:** Godi Bangla, Nova Cuttak, Sanchi and Birkoli.
- **Tamil Nadu:** Pachai Kodi and Vellaikodi.
- **Uttar Pradesh:** Deswari, Kapoori, Maghai and Bangla.
- **West Bengal:** Bangla, Sanchi, Mitha, Kali Bangla and Simurali Bangla.
- Betalvine is cultivated as mixed crop in arecanut plantations drainage channels are made between every two, three or four rows of arecanut depending of soil type.

Propagation: The vines are propagated by terminal stem cutting or setts about 30-45 cm long.

- Setts obtained from the top portions of the vines are easy to root and hence best for planting.
- On an average 1,00,000 setts are required for planting one hectare.

Planting Season: November - December and January – February are optimum.

- **Aftercare**

1. Thinning & pruning of -leaves
2. Sapfilling
3. Earming-up
4. Manuring
5. Irrigation and drainage

- Construction of Bareja (rectangular structures) for artificial support and shade Barejas are normally made on slightly slopy land, near to a source of irrigation at a higher level than the adjoining area. There must be a slope in all directions for a quick drainage of excess water. Barejas are nothing but rectangular structures made up of bamboo or jute sticks which are normally having a height of 2–2.5 meters. These rectangular structures are covered with thatching using coconut leaves or straw or such other materials.

- **Lowering of vine:-** When the vines reach a height of about 3nt they not only lose their vitality but also become difficult to harvest. They are often regulated by 'Lowering them'. It is a process in which vines are united from their support and their lower portion coiled and buried in the soil, leaving free only a few nodes, at the growing end.
- **Training and pruning:-** One month after the planting, young shoots start appearing and these young shoots are trained along the support and tied with them using banana fiber or jute fiber once in every 15 to 20 days.
- Best time of a lowering is summer season.
- Lowering of vine also, helps in producing more tillers and vigorous growth of vines.

Harvesting and Yield

Harvesting of leaves starts from 6 months to 18 months after planting.

- On an average 75 to 100 lakh leaves/ha/year are obtained.

Plant Protection

Pests

- Scale insects
- Mites (Sevvattai)
- Sooty mould (Aphids)
- Mealy bugs.

Diseases

- Phytophthora Wilt
- Bacterial leaf spot, blight and bacterial stem rot
- Anthracnose (Theechal)
- Powdery mildew.

Chapter 12

Multiple Choice Question

1. Tea require months from plantation to reach plucking is:

(a) 18-20 (b) 22-24

(c) 36-40 (d) 30-34

2. Fruit of coffee is:

(a) Pome (b) Drupe

(c) Berry (d) Aggregate

3. Coconut water is rich source of:

(a) Vit-B (b) Vit -B_2

(c) Vit -B_6 (d) Vit- C +Cynocobalamin

4. .... variety of coconut is used for tender water purpose:

(a) Banawali (b) Pratap

(c) Malayan (d) Gangabondam Green Dwarf

5. .... is important disorder of Rubber:

(a) Bleeding (b) Pink disease

(c) Brown blast (d) None

6. Coconut Development Board is located in

(a) Mumbai (b) Cochin

(c) Bangalore (d) Calicut

7. India ranks in natural rubber production is:

(a) 1st (b) 2nd

(c) 3rd (d) 4th

8. Coconuts take how many months from fertilization to mature nuts?

(a) 5-6 Months (b) 7-8 Months

(c) 8-10 Months (d) 11-12 Months

9. Coconut word drive from:-

(a) Latin (b) Greek

(c) Spanish (d) None

10. Monoluron present in coconut oil that has:

(a) Anti-diabetics (b) Anti-Cancer

(c) Anti-HIV (d) All of these

11. Dwarf coconut pollination behavior is:-

(a) Self-pollinated (b) Cross-pollinated

(c) Often cross-pollinated (d) All of these

12. Tall coconut pollination behavior is:

(a) Self-pollinated (b) Cross-pollinated

(c) Often cross-pollinated (d) All of these

13. Coconut fully mature nut has how much coir percent?

(a) 10-20% (b) 20-30%

(c) 30-40% (d) 40-50%

14. Coconut Fruit is :

(a) Drupe (b) Berry

(c) Single Seeded Berry (d) Single Seeded Drupe

15. Which State have maximum productivity of coconut?

(a) Kerala (20621 nut/ha) (b) Tamil Nadu (20621 nut/ha)

(c) Karnataka (20621 nut/ha) (d) Maharashtra (20621 nut/ha)

16. Optimum temperature for coconut growth:

(a) 24°C (b) 27°C

(c) 32°C (d) 35°C

17. Coconut is a:

(a) Heliotropic plant (b) Sciophytes plant
(c) Xerophytic plant (d) Lithophytic plant

18. pH of coconut water is:

(a) 3.5-4.5 (b) 4.8-5.3
(c) 5.5-6.3 (d) 6-7

19. How much take time by coconut for flowering to ripening?

(a) 10-11 months (b) 9-11 months
(c) 11-12 months (d) 7-9 months

20. How much old seedling should be planted in coconut?

(a) 9-12 months (b) 18-20 months
(c) 12-15 months (d) 7-9 months

21. How much coconut uses in raw consumption?

(a) 40% (b) 55%
(c) 65% (d) 80%

22. In which year coconut first hybrid variety released?

(a) 1921 (b) 1927
(c) 1948 (d) 1932

23. Which stimulating agent present in Arecanut?

(a) Thein (b) Theol
(c) Tannin (d) Arecotin

24. Which stimulating agent present in Tea?

(a) Thein (b) Theol
(c) Tannin (d) Arecotine

25. Aroma in Tea due to:-

(a) Thein (b) Theol
(c) Tannin (d) Arecotine

26. Bitterness in Tea due to:

(a) Thein (b) Theol

(c) Tannin (d) Arecotine

27. Colour of tea is due to:-

(a) Theaflavins (b) Thearufigens

(c) a & b Both (d) Theol

28. Arecotine present in arecanut:-

(a) 0.1% (b) 0.2%

(c) 1% (d) 5%

29. Which compound is responsible for astringent taste of nuts?

(a) Polyphenol (b) Tanins

(c) a & b Both (d) Arecotine

30. *Nuli* is made from:-

(a) Green arecanut (b) Dried ripe arecanut

(c) Over ripe arecanut (d) Tender Fruit of arecanut

31. Kalipak is prepared from:-

(a) Green arecanut (b) Dried ripe Arecanut

(c) Over ripe arecanut (d) Tender fruit of arecanut

32. Kottapak/Chali is prepared from:-

(a) Green arecanut (b) Dried ripe arecanut

(c) Over ripe arecanut (d) Tender fruit of arecanut

33. India's rank in production of Arecanut:

(a) 1st (b) 2nd

(c) 3rd (d) 4th

34. India's rank in consumption of Arecanut:

(a) 1st (b) 2nd

(c) 3rd (d) 4th

35. How much take time by arecanut for flowering to ripening?

(a) 10-11 months (b) 9-11 months
(c) 11-12 months (d) 7-9 months

36. India's rank in production of Tea:

(a) 1st (b) 2nd
(c) 3rd (d) 4th

37. India's rank in consumption of Tea:

(a) 1st (b) 2nd
(c) 3rd (d) 4th

38. India's rank in export of Tea:

(a) 1st (b) 2nd
(c) 3rd (d) 4th

39. Which country has 1st rank in tea export in world wide?

(a) India (b) Sri Lanka
(c) China (d) Myanmar

40. Darjeeling Tea is known for:

(a) Cup character (b) Muscate flavour
(c) Taste flavour (d) All of these

41. Assam Tea is known for:

(a) Cup character (b) Muscate flavour
(c) Taste flavour (d) All of these

42. Nilgiris Tea is known for:

(a) Cup character (b) Muscate flavour
(c) Taste flavour (d) All of these

43. Light strength tea is prepared by:-

(a) Orthodox method (b) CTC method
(c) Both (d) None

44. Strong strength tea is prepared by:-
    (a) Orthodox method (b) CTC method
    (c) Both (d) None
45. Tea is:
    (a) Acidifuge (b) Epiphytes
    (c) Calcifuge (d) All of these
46. Severe most pruning of tea is called as:
    (a) Skiffing (b) Tipping
    (c) Thumb pruning (d) Collar pruning
47. It is also known as "Monoculture", *i.e.* single crop grown over a large area.
    (a) Commercial grain farming (b) Plantation farming
    (c) Multiple farming (d) Mixed farming
48. Light most pruning of tea is called as:
    (a) Skiffing (b) Tipping
    (c) Thumb pruning (d) Collar pruning
49. Best harvesting time of tea is:
    (a) Two leaves and two bud stage (b) Two leaves and a bud stage
    (c) One leaves anda bud stage (d) Two leaves and a bud stage
50. The 1st plucking of recovering bushes is called:
    (a) Skiffing (b) Tipping
    (c) Thumb pruning (d) Collar pruning
51. Most suitable irrigation method for tea is:
    (a) Flood method (b) Drip method
    (c) Sprinkler method (d) Basin method
52. Most of tea garden are located at altitude ranging from:-
    (a) 800-1000 m (b) 1000-1200 m
    (c) 1400-1600 m (d) More than 1600 m

53. Processed tea brightness is due to:

(a) Aflavine (b) Arubigins

(c) a & b Both (d) Withering

54. Withering is done for:

(a) Increasing of fermentation (b) Taste enhancement

(c) Brightness (d) Removing of moisture

55. Processed tea has moisture content:

(a) 1.5-2% (b) 2.5-3%

(c) 3-5% (d) 5-7%

56. Largest tea producer state of India is:

(a) Kerala (b) Assam

(c) Arunachal Pradesh (d) West Bengal

57. Which rank in coffee production of India in world?

(a) 1st (b) 4th

(c) 6th (d) 8th

58. Leaf Rust resistant coffee is:

(a) *Coffea robusta* (b) *Coffea arabica*

(c) *Coffea liberica* (d) All of theses

59. In world, consumption ratio of natural and synthetic rubber is:

(a) 60:40 (b) 39:61

(c) 70:30 (d) 45:55

60. India's rank in rubber production is:

(a) 1st (b) 2nd

(c) 3rd (d) 5th

61. Niacin which is useful to cure skin diseases, is present:

(a) Rubber (b) Tea

(c) Coconut (d) Coffee

62. Scuffling is practiced in which crop?
    (a) Rubber (b) Tea
    (c) Coconut (d) Coffee
63. Vitamin-A content in Betelvine is:
    (a) 4800 IU (b) 9600 IU
    (c) 2020 IU (d) 6200 IU
64. Which training method is commercial used in coffee?
    (a) Single stem (b) Multiple stem system
    (c) Double stem (d) All of these
65. Which variety of coffee is growing at large scale in India?
    (a) Kent (b) S-795
    (c) Agro (d) All of these
66. Cocoa horizontal branches are called:-
    (a) Chupan (b) Fan or Jorquette
    (c) Janams (d) All of theses
67. Cocoa vertical stem are called:
    (a) Chupan (b) Fan or Jorquette
    (c) *Janam*s (d) all of tham
68. Which is useful for measurement of rubber percent ?
    (a) Refractometer (b) Monometer
    (c) Matrola (d) All of them
69. IIHR is established at:
    (a) Bangalore (b) New Delhi
    (c) Lucknow (d) Jodhpur
70. The moisture content of dried cashew nuts should be
    (a) Nearly 2% (b) Nearly 5%
    (c) Nearly 8% (d) Nearly 12%

71. Which one of the following fruits was introduced in India for soil conservation purpose but later on it attained the status of a commercial fruit crop?

(a) Cashew nut (b) Sapota

(c) Date palm (d) Arecanut

72. The moisture content of 'desiccated coconut' should be:-

(a) Less than 1% (b) Less than 3%

(c) Less than 5% (d) Less than 6.5%

73. Tea mosquito bug is a common pest of:-

(a) Cashew (b) Tea

(c) Coffee (d) Rubber

74. Coconut water contains.........% of water:-

(a) 94 (b) 94.5

(c) 96 (d) 96.5

75. Optimum tapping depth in rubber should be:

(a) 1 cm (b) 1 mm

(c) 0.1 mm (d) 0.01 cm

76. Coconut water contain which hormone:

(a) Auxin (b) GA_3

(c) Cytokinin (d) TIBA

77. Varieties of betel vine are:

(a) Karpurakodi, Kallarkodi, Revesi, Karpuri

(b) SG 1, SG (BV) - 2, Vellaikodi, Pachaikodi

(c) Sirugamani 1, Anthiyur kodi, Kanyur kodi, Bangla

(d) All of these

78. Betel vine is commercially propagated by:

(a) Terminal stem cutting (Setts) (b) Seed

(c) Layering (d) Runners

79. Best time of betel vine plantation is

(a) November - December	(b) January - February
(c) a & b Both	(d) None

80. Yield of betel vine per hector is:

(a) 75 to 100 lakh leaves/ha/year	(b) 50 to 70 lakh leaves/ha/year
(c) 100-125 lakh leaves/ha/year	(d) 125-150 lakh leaves/ha/year

81. Betel vine is:

(a) Monocious	(b) Dioecious
(c) Complete flower	(d) Gaynodioeucious

82. Commercial cultivation of rubber in India was started in

(a) 1902	(b) 1909
(c) 1915	(d) 1947

83. *Hevea* seeds normally ripen during:- July-September

(a) Nov.-Dec.	(b) July-September
(c) Feb-March	(d) May-June

84. Which is commercial propagation method of rubber :-

(a) T-budding	(b) Brown budding
(c) Cutting	(d) Patch budding

85. Studies on tissue culture of rubber plants were started in :

(a) 1902	(b) 1909
(c) 1966	(d) 1947

86. Tissue culture plant of rubber mostly prepared from:-

(a) TNAU	(b) RRII
(c) CPRI	(d) IARI

87. For obtaining optimum yield, it is preferable to consume bark annually is:

(a) 20-23 cm	(b) 17-18 cm
(c) 14-16 cm	(d) 12-14 cm

88. Which Tapping frequency is best or optimum yield of rubber?

(a) d2 (b) d3

(c) d4 (d) d6

89. In which tapping frequency done maximum tap per year:-

(a) d2 (b) d3

(c) d4 (d) d6

90. How many tap done in a year for d2 tapping frequency?

(a) 50 (b) 100

(c) 150 (d) 200

91. Tapping cut in rubber of the budded trees should have a slope of about

(a) 30^0 and horizontal (b) 30^0 and vertical

(c) 25°C and horizontal (d) 25°C and vertical

92. Tapping cut in rubber of the seedling trees should have a slope of about

(a) 30^0 and horizontal (b) 30^0 and vertical

(c) 25°C and horizontal (d) 25°C and vertical

93. *Hevea* latex in the latex vessels of tapped trees how much contains rubber in the form of particles:

(a) 35-45% (b) 45-55%

(c) 25-35 % (d) 55-65%

94. *Hevea* in binding is most efficient in:

(a) Saline pH (b) Natural pH

(c) Acidic pH (d) All of these

95. The number of trees allotted to a tapper for a day's tapping is known as :

(a) Tapping (b) Tapping task

(c) Tapping duration (d) None of theses

96. In India tapping task is:

(a) 100-200 (b) 200-300

(c) 300-400 (d) 400-500

97. In India most suitable tapping system in budded tree of rubber is:

(a) One full spiral alternate daily (S d2)

(b) Half spiral third daily (S/2 d3)

(c) Half spiral daily daily (S/2 d21)

(d) Half spiral alternate daily (S/2 d2)

98. What is term use for the area of bark in which tapping cut is located?

(a) Tapping cut (b) Tapping area

(c) Tapping position (d) Panel

99. Which type of tapping done in old rubber tree:

(a) High Level Tapping (b) Intensive Tapping

(c) Controlled Upward Tapping (d) Low Frequency Tapping (LFT)

100. Which tapping method done for longer exploitation of the virgin bark above the basal panel is?

(a) High Level Tapping (b) Intensive Tapping

(c) Controlled Upward Tapping (d) Low Frequency Tapping (LFT)

101. Which chemical used for increasing of rubber production?

(a) Ethylene (b) Chloride ions

(c) Auxion (d) A & B both

102. How much concentration used of ethylene for increasing of rubber production?

(a) 1.5% (b) 2.5%

(c) 5% (d) 10%

103. Highly intensive tapping prior to replanting is the last stage in the tapping cycle is called:-

(a) Intensive Tapping (b) Controlled Upward Tapping

(c) High Level Tapping (d) Slaughter Tapping

104. Economic life of rubber tree is:

(a) 10 year (b) 20 year

(c) 30 year (d) 40 year

105. Specific gravity of rubber latex is:

(a) 1.021-1.036 (b) 1.102-1.205

(c) 0.725-0.825 (d) 0.974 and 0.986

106. How much water content in rubber latex?

(a) 30-40% (b) 55-65%

(c) 40-55% (d) 80-85%

107. Which of following chemical used for preservation of rubber latex?

(a) Acetic acid (b) Ammonia

(c) Sulphuric acid (d) Malice acid

108. Which is Concentration creaming agent is?

(a) Ammonium alginate (b) Tamarind seed powder

(c) B&C (d) None of these

109. How much dry rubber content in concentration by creaming?

(a) 40-50% (b) 50-55%

(c) 55-60% (d) 60-65%

110. How much dry rubber content in concentration by centrifugation?

(a) 40-50% (b) 50-55%

(c) 55-60% (d) 60-65%

111. Skim latex is generally coagulated by:-

(a) Formic acid (b) Acetic acid

(c) Sulphuric acid (d) All of these

112. Small sand dirt, or foreign particles in the rubber sheet is:-

(a) Improper bulking (b) Improper sieving

(c) Insufficient mixing of the acid (d) a & b

113. Small bubbles along the edges of rubber sheets is due to:-

(a) Insufficient acid for coagulation

(b) Insufficient mixing of the acid with latex

(c) Improper sieving (d) a & b

114. Pinhead bubbles in clusters all over the sheets is due to:-

(a) Insufficient acid for coagulation

(b) Insufficient mixing of the acid with latex

(c) Bacterial growth

(d) a & b

115. Small white specks and irregular bubbles in rubber sheet is due to:-

(a) Precoagulated latex

(b) Rapid drying in the smoke house

(c) Bacterial growth

(d) All of these

116. Blisters and large Bubbles in rubber sheet is due to:-

(a) Precoagulated latex
(b) Rapid drying in the smoke house

(c) Bacterial growth
(d) All of these

117. Burnt and oxidized sheets in rubber sheet is due to:-

(a) High temperature drying
(b) Flame directly reaching the sheet

(c) Improper sieving
(d) a & b

118. Mould growth in rubber sheet is due to:-

(a) Improper drying

(b) Storage under moist conditions

(c) Improper washing of the coagulum during sieving

(d) a & b

119. Rust growth in rubber sheet is due to:-

(a) Improper washing of the coagulum during and after sheeting

(b) Storage under moist conditions

(c) Improper washing of the coagulum during sieving

(d) a & b

120. Stickiness in rubber sheet is due to:

(a) High dosage of coagulant
(b) High temperature drying

(c) Drying in sunlight alone
(d) All of these

121. Discolouration in rubber sheet is due to:

(a) Atmospheric oxidation during coagulation

(b) High dosage of coagulant

(c) High temperature drying

(d) Drying in sunlight alone

122. Rubber is mostly used in:

(a) Tyres and tubes (b) Battery

(c) Foam Mattresses (d) Balloons

123. Seed harvesting in rubber is during:-

(a) April-June (b) July-September

(c) October-November (d) January-March

124. Rubber seed cake used in:

(a) Cattle feed (b) Paltry feed

(c) Fertilizer (d) a & b

125. Suitable pH for coconut cultivation?

(a) 5.5-7.5 (b) 6.5-7.5

(c) 7.2-8.0 (d) 7-9.5

126. Planting distance for coconut planting is:

(a) 5 X 5 m (b) 6.5 X 6.5 m

(c) 7.5 X 7.5 m (d) 10X10 m

127. How much coconut plant recommend for planting of one hector?

(a) 100 (b) 157

(c) 166 (d) 177

128. Seedling age of coconut should be for plantation is:

(a) 12-14 months (b) 14-18 months

(c) 9-12 months (d) 9-12 months

129. Ideal time for coconut planting is:

(a) May (b) June

(c) July (d) August

130. How much time taken by coconut become mature from opening of spathe?

(a) 7-8 months (b) 8-9 months

(c) 9-10 months (d) 11-12 months

131. Economic life of the coconut palm is about :

(a) 50 years (b) 60 years

(c) 70 years (d) 80 years

132. How much oil content in coconut?

(a) 55% (b) 60%

(c) 50% (d) 65%

133. Which state is largest producer of tea:

(a) Assam (b) Kerala

(c) Karnataka (d) Tamil Nadu

134. Which country is largest producer of tea:

(a) Brazil (b) India

(c) Sri lanka (d) China

135. Which country is largest producer of coffee:

(a) Brazil (b) India

(c) Sri Lanka (d) China

136. India rank in coffee production is:

(a) 1st (b) 3rd

(c) 4th (d) 6th

137. Rhinoceros beetle is important insect of:

(a) Cashew nut (b) Coffee

(c) Rubber (d) Coconut

138. Planting distance of cashew nut is:

(a) 5X5 m (b) 7X7 m

(c) 9X9 m (d) 10X10 m

139. Yield of cashew nut is:

(a) 1-2Kg/tree (b) 3-4Kg/tree
(c) 5-6Kg/tree (d) 7-8 Kg/tree

140. Cashew nut is belong to group of:

(a) Papaya (b) Clove
(c) Mango (d) Palmyra palm

141. Cashew nut is:

(a) Self-pollinated (b) Cross-pollinated
(c) Often-cross pollinated (d) All of these

142. Best method of cashew propagation:

(a) T-Budding (b) Tongue Grafting
(c) Side Grafting (d) Seed

143. Cashew seedlings are grown under shade % is:

(a) 50% (b) 30%
(c) 45% (d) 65%

144. Cashew nut is:

(a) Monoecious (b) Dioecious
(c) Androdioecious (d) Gynodioecious

145. Origin of cashewnut is:

(a) Brazil (b) India
(c) China (d) South-Africa

146. Fat percent in cashewnut is:

(a) 64.5% (b) 58.5%
(c) 46% (d) 46.8%

147. Drupe fruits are developed from............ovary.

(a) Inferior (b) Superior
(c) Epigynous (d) Hypogynous

148. Which state is highest productivity of cashewnut?

(a) Kerala (b) Maharashtra

(c)Karnataka (d) Tamil Nadu

149. Which State is highest producer of cashew nut?

(a) Kerala (b) Maharashtra

(c) Karnataka (d) Tamil Nadu

150. Pruning of cashew nut should be carried out in month:

(a) May-June (b) January-February

(c) November-December (d) August- September

151. Kernels are dried to .... % moisture content before they are packed.

(a) 2% (b) 3%

(c) 4% (d) 7%

152. Which is largest exporter country of cashew nut?

(a) India (b) Brazil

(c) China (d) Thailand

153. India share......................% of cashew nut export of total world export.

(a) 42% (b) 53%

(c) 65% (d) 75%

154. How much protein content in cashew nut?

(a) 15.6% (b) 18.5%

(c) 20.88% (d) 21.20%

155. Cashew nut introduce during .... century in India.

(a) 14 (b) 12

(c) 19 (d) 16

156. Who introduce the cashew nut in India?

(a) Portuguese (b) British

(c) Mogul (d) Turkish

157. Cashew can be distilled to produce alcoholic drink ....

(a) Perry (b) Cider

(c) Eau de vie (d) Fenny

158. Average weight of cashew nut is:

(a) 5-7g (b) 7-10g

(c) 11-14g (d) 15-18g

159. Which method is best method of cashew nut roasting?

(a) Drum roasting (b) Steam roasting

(c) Oil bath roasting (d) All of these

160. Which method is widely used for cashew nut roasting?

(a) Drum roasting (b) Steam roasting

(c) Oil bath roasting (d) All of these

161. Removal of testa (seed coat) from the kernel is called:

(a) Shelling (b) Peeling

(c) Roasting (d) Testing

162. Cashew apple is rich source of:

(a) Vit-A (b) Vit-B

(c) Vit-C (d) Vit-E

163. Who much content of sucrose of palmyra palm jaggery?

(a) 50% (b) 60%

(c) 68% (d) 79%

164. The chief product of the palmyra is the sweet sap obtained by tapping the tip of the inflorescence is called :

(a) Toddy (b) Palm wine

(c) Liquor (d) Jaggery

165. New Guinea Palmyra palm is:

(a) *Borassus aethiopum* (b) *Borassus akeassii*

(c) *Borassus flabellifer* (d) *Borassus heineanus*

166. African Palmyra palm is:

(a) *Borassus aethiopum* (b) *Borassus akeassii*

(c) *Borassus flabellifer* (d) *Borassus madagascariensis*

167. Sambirano Palmyra palm is:

(a) *Borassus aethiopum* (b) *Borassus sambiranensis*

(c) *Borassus flabellifer* (d) *Borassus madagascariensis*

168. Sutable rainfall for tea cultivation is:-

(a) 100 cm (b) 150-300 cm

(c) 300-400 cm (d) 500cm

169. Which crop share is in highest production of plantation crop

(a) Tea (b) Cashew nut

(c) Coconut (d) Coffee

170. Which state is highest in productivity of coconut?

(a) Kerala (b) Lakshadweep

(c) Karnataka (d) Andhra Pradesh

171. Yellow leaf is a serious diseases of……………….

(a) Cocoa (b) Arecanut

(c) Cashew nut (d) Coconut

172. Which of the following has maximum oil content?

(a) Oil palm (b) Groundnut

(c) Mustard (d) Linseed

173. Root wilt is serious diseases of:

(a) Oil palm (b) Cashewnut

(c) Arecanut (d) Coconut

174. Coconut development Board is located at:

(a) Bangalore (b) New Delhi

(c) Cochin (d) Hyderabad

175. Immature coconut fruit are harvested after:

(a) 5-7 months (b) 8-10 months

(c) 10-12 months (d) 12-14 months

176. Economic part of tea plants which are harvested:

(a) Bud & young leaves (b) Mature leaves

(c) Stem (d) Bark

177. Which of the following is not a variety of arecanut?

(a) Mangala (b) Sumangala

(c) Kaddyam (d) Sreemangala

178. Coconut Development Board works under the:

(a) ICAR (b) Govt. of India

(c) State Govt. (d) None of these

179. Commercial propagation method of tea is:

(a) Seed (b) Layering

(c) Budding (d) Cutting

180. Which of the following is having maximum oil content?

(a) Groundnut (b) Coconut

(c) Oil Palm (d) Soybean

181. Stimulent which is present in tea:

(a) Caffein (b) Piperine

(c) Thiobromin (d) Atropine

182. Origin center of Cocoa is :

(a) Asia (b) Africa

(c) Brazil (d) Africa

183. In world best quality tea is grown at:

(a) China (b) Nilgiri (India)

(c) Brazil (d) Darjeeling (India)

184. Cashew nut is commercially propagated through:

(a) Cutting (b) Patch budding

(c) Softwood grafting (d) Layering

185. Which of the following comes under narcotic drug:

(a) Tea (b) Coffee

(c) Palmyra palm (d) Poppy

186. Major prevailing problem in coconut cultivation:

(a) Lack of good variety (b) Red palm weevil

(c) Stem bleeding (d) Eryophite mite

187. Natural rubber is mostly used in which of the following industry:

(a) Sheet (b) Box

(c) Tyres (d) All of these

188. Red rust of tea is caused by:

(a) Algae (b) Virus

(c) Fungi (d) Bacteria

189. Identify a beverage crop in the following:

(a) Lemon (b) Coconut

(c) Lime (d) Guava

190. Identify the crop plant bearing cauliflorus bud:

(a) Cocoa (b) Coconut

(c) Banana (d) Pineapple

191. Which one of the following dose not bear mixed bud?

(a) Apple (b) Pear

(c) Mango (d) Coconut

192. Name the plant which is homogamous?

(a) Mango (b) Cashew nut

(c) Coconut (d) Pea

193. Identify the plantation crop which yield maximum oil per unit area:

(a) Coconut (b) Olive
(c) Oil Palm (d) Palmyra Palm

194. Betelvine is a:

(a) Hermaphrodite (b) Dioecious
(c) Andromonocious (d) Monocious

195. Plantation crops occupied of total crop land is

(a) 3.82% (b) 2%
(c) 4.5% (d) 6%

196. Total earnings of plantation crops from the export of agricultural produces:

(a) 55% (b) 65%
(c) 75% (d) 85%

197. Which one of the following dose not variety of betel vine?

(a) Pachaikodi (b) Sirugamani 1
(c) Anthiyur kodi (d) Kavery

198. For propagation of betel vine cutting should have how many nodes?

(a) 3-5 (b) 4-7
(c) 7-10 (d) 1-2

199. Suitable time of betel vine plantation in open field?

(a) June- July (b) September
(c) October (d) Feb-March

200. Cocoa world drive from which world?

(a) Latin (b) Greek
(c) Spinach (d) English

201. Which type of cocoa is occupied in largest area?

(a) Forastero (b) Criollo
(c) Trinitario (d) All of these

202. Coffee fruit is called as:

(a) Drupe (b) Pea berry
(c) Nut (d) Pome

203. Among following which is dioecious:

(a) Date palm (b) Palmyrha
(c) Both (d) Coconut

204. Native of coffee is:

(a) Indonesia (b) China
(c) Ethiopia (d) Australia

205. Chintamani-1 is the variety of:

(a) Cashew nut (b) Coconut
(c) Mango (d) Cocoa

206. Coffee board was established in the year:

(a) 1974 (b) 1987
(c) 1942 (d) 1992

207. Which one of following is called Ice apple?

(a) Coconut (b) Arecanut
(c) Cocoa (d) Palmyra palm

208. Betel vine is good source of:

(a) Vitamin-A (b) Calcium
(c) Both (d) Iron

209. Oil content of betel vine leaf is:

(a) 0.7 to 2.6% (b) 2.5-3.5%
(c) 0.1% (d) 0.4%

210. Chandan is variety of:

(a) Arecanut (b) Cocoa
(c) Palmyra Palm (d) Betel vine

211. Sapfilling is done in:

(a) Arecanut (b) Cocoa

(c) Palmyra Palm (d) Betel vine

212. Manufacture of milk chocolate was discovered by:

(a) J. Smith, 1958 (b) M.D.Peter, 1876

(c) R.T.Mayer, 1888 (d) Wolf, 1952

213. Which vitamin present in cocoa shell?

(a) Vit-A (b) Vit-D_1

(c) Vit-D_7 (d) Vit-B_{12}

214. India share ....% arecanut production of total world production.

(a) 47 (b) 66

(c) 77 (d) 52

215. India contributes of total world production:

(a) 47 % (b) 66%

(c) 77% (d) 52%

216. Coconut today obtained from:

(a) Fruit (b) Inflorescences

(c) Stem (d) Root

217. Cashew nut is rich source of:

(a) Vit-B_1 (b) Vit-B_2

(c) Vit-B_6 (d) Vit-B_7

218. In cashew nut blossoming takes place in:

(a) January-March (b) June-July

(c) October-November (d) November- January

219. Generic name of oil palm "*Elaeis*" drive from:

(a) Latin (b) Greek

(c) English (d) Spanish

220. How much oil palm plant recommend per hector?

(a) 157 (b) 177

(c) 143 (d) 100

221. Removal of such inflorescences in oil palm during initially stage is called:

(a) Ablation (b) Pruning

(c) Disbudding (d) Pinching

222. Pruning time in oil palm is:

(a) February-March (b) June-July

(c) August- September (d) November- December

223. Oil palm bunches usually take time for ripen after anthesis about:

(a) 5 months (b) 6 months

(c) 9 months (d) 12 months

224. Youngest opened leaves of tea are picked is called:

(a) Flower Pekoe (b) Souchong

(c) Orange Pekoe (d) None of these

225. Tiny shoots and unopened of tea are picked is called:

(a) Flower Pekoe (b) Souchong

(c) Orange Pekoe (d) None of these

226. Older, coarser leaves closer to the trunk of tea are picked is called:

(a) Flower Pekoe (b) Souchong

(c) Orange Pekoe (d) None of these

227. Partially fermented tea leaf is called:

(a) Green Tea (b) Black Tea

(c) Oolong Teas (d) Silver tea

228. India share of coffee export in world is:

(a) 2.05% (b) 9.08%

(c) 4.04% (d) 12.05%

229. Wider adaptability of Arabica coffee varietyin India is:
   (a) S.795 (b) Sln.5-B
   (c) Sln.6 (d) Sln.9

230. Suitable planting time of coffee is:
   (a) February-March (b) November-December
   (c) May-June (d) August-September

231. Which of following coffee recommend double topping?
   (a) Tall arabica (b) Dwarf arabica
   (c) Robustas (d) b&c

232. Which of following coffee recommend single topping?
   (a) Tall arabica (b) Dwarf arabica
   (c) Robustas (d) b&c

233. India has an area under cashew nut cultivation is (2010-11):
   (a) 7.60lakh ha (b) 9.53 lakh ha
   (c) 10.20 lakh ha (d) 5.08 lakh ha

234. Recommend plantation distance of cashew nut is:
   (a) 5x5 m (b) 6x6 m
   (c) 7.5x7.5 m (d) 9x9 m

235. Pit size for cashew nut planting is:
   (a) 60cm x 60cm x 60cm (b) 50cm x 50cm x 50cm
   (c) 75cm x 75cm x 75cm (d) 1m x1m x 1m

236. Pruning time of cashew nut is:
   (a) February-March (b) November-December
   (c) May-June (d) August-September

237. The shelling percentage of cashew nut is:
   (a) 33% (b) 40%
   (c) 20-25% (d) 40-50%

238. Breaking cashew kernel is called as:

(a) Whole (b) Splits

(c) Scorched kernels (d) None of theses

239. The average productivity of coffee in India is:

(a) 1653kg/ha (b) 1052 kg/ha

(c) 792 kg/ha (d) 908 kg/ha

240. Which of following country has highest productivity of coffee in the world?

(a) India (b) Guinea

(c) Brazil (d) Costa Rica

241. Kerala's is account area and production of cocoa is:

(a) 76% & 78% (b) 78% & 76%

(c) 82% & 85% (d) 50% & 60%

242. Which one is associated with tea research?

(a) TNAU (b) Tocklai

(c) IARI (d) CISH

243. Which one is calcifiuge crop?

(a) Oil Palm (b) Betel vine

(c) Arecanut (d) Tea

244. Tea take time to reach plucking from plantation is:

(a) 24 months (b) 26-28 months

(c) 12-16 months (d) 18-20 months

245. In which crop useful of iodine test for estimation the starch content in root before pruning?

(a) Coconut (b) Cashew nut

(c) Tea (d) Rubber

246. Which pruning method used in coffee?

(a) Double stem open center stem (b) Single stem open center stem

(c) Double stem leader center stem (d) Single stem leader center stem

247. Which one of following is tree coffee?

(a) *Coffea liberica* (b) *C.canephora*

(c) *C.arbica* (d) None of these

248. Coffee fruit with single seeded is called:

(a) Triangular berry (b) Normal berry

(c) Pea berry (d) Elephant berry

249. Which of following is agent of coffee pollination?

(a) Wind (b) Bee

(c) Both (d) House fly

250. Which one is high yielding clone of tea?

(a) Singara (b) TTRI-1

(c) TTRI-2 (d) Sundaram

251. Coffee pea berry formed by:

(a) Non-fertilization (b) Abortion of embryo

(c) Both (d) Normal fertilization

252. Which one is high yielding clone of tea?

(a) Singara (b) TTRI-1

(c) TTRI-2 (d) Sundaram

253. Cauvery is variety of coffee developed by cross between:

(a) Catimor x Caturra (b) Caturra x Hi brido Timor

(c) Devamachy x S -881 (d) S.288 x Kent

254. Which one is the popular clone is used as male parent in coffee breeding program?

(a) Taferikela (b) Cattura

(c) Catimor (d) Kents

255. Fat content in cocoa is:

(a) 47% (b) 63%

(d) 21% (d) 37%

256. Cocoa bearing habit is:

(a) Cladophyll (b) Climbing

(c) Cauliflora (d) All of theses

257. Which one is suitable for commercial cocoa cultivation?

(a) Trinitarion (b) Criollo

(c) Forestero (d) All of theses

258. Cashew apple is:

(a) Drupaceous nut (b) Fleshy Peduncle

(c) Seed (d) Epicarp

259. Which is suitable position of seed sowing is:

(a) Horizontal (b) Vertical

(c) Both (d) None of theses

260. Edible portion of coconut is:

(a) Mesocarp (b) Endocarp

(c) Placenta (d) Endosperm

261. Which of flowing is autogamous?

(a) Dwarf coconut (b) Tall coconut

(c) a & b Both (d) None of theses

262. Crown choking in coconut is due to:

(a) Ca deficiency (b) Mg deficiency

(c) Fe deficiency (d) B deficiency

263. Rossetee and Little leaf is due to:

(a) Ca deficiency (b) Mg deficiency

(c) Zn deficiency (d) B deficiency

264. Which one of the following is suitable arecanut variety for Andaman Island?

(a) Cal-9 (b) Cal-17

(c) Cal-21 (d) Cal-10

265. Tapping in rubber should be done in:

(a) Afternoon (b) Evening

(c) Morning (d) All of these

266. Astringency in areanut is due to:

(a) Polyphenols (b) Tannins

(c) a & b Both (d) Arecoline

267. Which one of following is high yielding clone of rubber?

(a) RRII-105 (b) RRII-110

(c) Tjir-1 (d) Tjir-2

268. Which of following is called "Small holders irrigated crop"?

(a) Coconut (b) Cashew nut

(c) Betel vine (d) Oil palm

269. The shell covering the kernel inside the fruit of oil palm is called?

(a) Dura (b) Pisifera

(c) Tanera (d) All of these

270. Suitable temperature for oil palm seed germination is:

(a) 20-25°C (b) 25-30°C

(c) 30-35°C (d) 39-40°C

271. Pollination agent in oil palm is:

(a) Wind (b) Weevil

(c) House fly (d) Bee

272. Oil palm mesocarp oil content is:

(a) 35-45% (b) 45-55%

(c) 55-60% (d) 65%

273. Coffee rust is caused by:

(a) *Uromyces hobsoni* (b) *Hemileia vastatrix*

(c) *Cronatium ribicola* (d) *Ravenelia embliccae*

274. Oil palm was introduced as an ornamental crop at National Botanical Garden, Kolkatta during:

(a) 1860 (b) 1960

(c) 1848 (d) 1840

275. Shade tree are planted in tea area mainly for:

(a) Enhancing rainfall (b) Lowering frost occurrence

(c) Improving fertility of soil (d) Minimizing temperature

276. The optimum temperature (°C) for growing tea is:-

(a) 15-22 (b) 28-32

(c) 20-27 (d) 32-38

277. Cashew nut export is rising at the rate of:

(a) 33% (b) 22%

(c) 11% (d) 40%

278. Coffee is:

(a) LDP (b) DNP

(c) MDP (d) SDP

279. Coconut variety having un-branched inflorescence:

(a) Dwarf variety (b) Tall variety

(c) Spikata variety (d) All of these

280. Average plant density maintained in coconut is:

(a) 177 (b) 156

(c) 123 (d) 100

281. Black pod is diseases of:

(a) Coconut (b) Betel vine

(c) Cocoa (d) Coffee

282. Once the oil palm bunches are harvested they should be processed within:

(a) 60 hours (b) 48 hours

(c) 36 hours (d) 24 hours

283. Penducle and cotyledons are edible part in fruit of:

(a) Cashew nut (b) Cocoa

(c) Arecanut (d) Coconut

284. An example of polygamous monoecious plant is:

(a) Coconut (b) Cashew nut

(c) Arecanut (d) Tea

285. Arecanut seed has:

(a) Embryo dormancy (b) Seed coat dormancy

(c) Rest period (d) Chemical dormancy

286. Yellow leaf disease is serious disease of:

(a) Cocoa (b) Coconut

(c) Tea (d) Arecanut

287. Anab-e-roga is a serious diseases of:

(a) Tea (b) Coffee

(c) Coconut (d) Rubber

288. Palm oil is obtained from:

(a) Endocarp (b) Epicarp

(c) Mesocarp (d) None of these

289. Majority of coconut roots are concentrated within a diameter of:

(a) 6m (b) 4 m

(c) 2 m (d) 8 m

290. Planting distance of arecanut is:

(a) 5x5 m (b) 3x3 m

(c) 2x2 m (d) 2.7x2.7 m

291. Which of following states is largest producer of plantation crops?

(a) Kerala (b) Karnataka

(c) Maharashtra (d) West Bengal

292. Which of following states is largest producer of plantation crops?

(a) Kerala (b) Karnataka

(c) Maharashtra (d) West Bengal

293. Cashew nut variety with bold nut is:

(a) Kanaka (b) Dhana

(c) HS-117 (d) H-1591

294. Horticulture society of India (HIS) was establish in:

(a) 1947 (b) 1942

(c) 1884 (d) 1960

295. Central Coffee Research Institute is located at:

(a) Balasur (b) New Delhi

(c) Trichy (d) Eluru

296. Red rust of tea is caused by:

(a) Virus (b) Fungi

(c) Bacteria (d) Algae

297. Identify the crop in the following:

(a) Pineapple (b) Coconut

(c) Banana (d) Cocoa

298. Mangala is a variety of :

(a) Arecanut (b) Cocoa

(c) Tea (d) Coffee

299. The coffee plant grow well in temperature range of:

(a) 0-10°C (b) 12-36°C

(c) 42-50°C (d) 36-42°C

300. The edible part of coconut and cashew nut is:

(a) Peduncle and Cotyledons (b) Endocarp and Mesocarp

(c) Endosperm, Peduncle and Cotyledon (d) Pericarp and Perianth

Chapter 13

Matching

	Crop	Highest producer state
301.	Arecanut	(a) Assam
302.	Cashew nut	(b) Kerala
303.	Rubber	(c) Andhra Pradesh
304.	Tea	(d) Karnataka

	Crop	Chromosome No.
305.	Betelvine	(a) 2n=2x=42
306.	Cashew nut	(b) 2n=2x=26
307.	Cocoa	(c) 2n=2x=22
308.	Coconut	(d) 2n = 2x = 32
309.	Arabica Coffee	(e) 2n=4x=44
310.	Robusta Coffe	(f) 2n=2x=36

	Name of Board, Directorates and Organizations	Establishment Year
311.	Coffee Board of India	(a) 1947
312.	Tea Board of India	(b) 1970
313.	Indian Rubber Board	(c) 1942
314.	Coconut Development Board	(d) 1st April 1954
315.	Central Plantation Crops Research Institute (CPCRI)	(e) 1979

	Crop	Leading country in area
316.	Cashew nut	(a) Indonesia
317.	Cocoa	(b) Malaysia
318.	Coffee	(c) India
319.	Oil Palm	(d) Thailand
320.	Rubber	(e) Brazil

	Crop	Fruit Type
321.	Arecanut	(a) Berry
322.	Coffee	(b) Globose berry
323.	Cashew nut	(c) One-seeded berry
324.	Coconut	(d) Droup
325.	Betel	(e) Droup

	Crop	Variety
326.	Arabica Coffee	(a) Ghanagate Bangla
327.	Cocoa	(b) Sel-2R (S-270)
328.	Betel vine	(c) Selection 1 (S288)
329.	Rubber	(d) CCRP 1
330.	Robusta Coffee	(e) RRII 105

	Crop	Variety
331.	Tall Coconut	(a) Vengurla-1
332.	Dwarf Coconut	(b) Lakshadweep Ordinary
333.	Tea	(c) Mohitnagar
334.	Areca nut	(d) Singara
335.	Cashew nut	(e) Udha Gangapani

	Crops	Useable Part
336.	Rubber	(a) Leaf
337.	Betelvine	(b) Bean
338.	Tea	(c) Latex
339.	Cocoa	(d) Endocarp
340.	Coconut	(e) Leaves/ shoots

	Name of Board, Directorates and Organizations	Head Quarter
341.	Coconut Development Board	(a) Kottayam, Kerala
342.	Central Plantation Crops Research Institute (CPCRI)	(b) Puttur, Karnataka
343.	Rubber Research Institute	(c) Pedavegi, AP
344.	Directorate of Cashew Research	(d) Kasargod, Kerala
345.	Directorate of oil Palm Research	(e) Kochi, Kerala

	Crop	Orizin
346.	Areca nut	(a) North East Brazil in LatinAmerica
347.	Cashew nut	(b) Central and South America.
348.	Cocoa	(c) India and South Eastern Countries.
349.	Coconut	(d) East and South Asia
350.	Tea	(e) South East Asia and Malaysia

	Crop	Family
351.	Betelvine	(a) Palmaceae
352.	Cashew nut	(b) Sterculiaceae
353.	Cocoa	(c) Anacardiaceae
354.	Coconut	(d) Rubiaceae
355.	Coffee	(e) Piperaceae

	Crop	Type of Inflorescence
356.	Betel vine	(a) Spadix
357.	Coffee	(b) Panicles
358.	Areca nut	(c) Dichasial
359.	Cocoa	(d) Cyme
360.	Cashew nut	(e) Spike

	Crops	Protein
361.	Betel vine	(a) 3.0%
362.	Cashew nut	(b) 3.1%
363.	Coconut	(c) 5%
364.	Cocoa	(d) 3.3%
365.	Areca nut	(e) 21.2%

	Crop	Commercial Propagation Method
366.	Betel vine	(a) Seed
367.	Rubber	(b) Hard Wood Cutting
368.	Coconut	(c) Side grafting
369.	Tea	(d) Terminal Stem Cutting
370.	Cashew nut	(e) Green budding

	Rubber	Botanical Name
371.	Cera Rubber	(a) *Parthenium argentatum*
372.	Para Rubber	(b) *Manihot glaziovii*
373.	Indian Rubber	(c) *Castiolla elastic*
374.	Panama Rubber	(d) *Hevea brasiliensis*
375.	Guayul Rubber	(e) *Ficus elastic*

	Crop	Important Practices
376.	Tea	(a) Blending
377.	Cashew nut	(b) Tapping
378.	Rubber	(c) Withering
379.	Coffee	(d) Digestion
380.	Oil Palm	(e) Roasting

	Crop	Physiological Disorder
381.	Coconut	(a) Nut-Splitting
382.	Coffee	(b) Orange spotting disease
383.	Rubber	(c) Hot and cold disease
384.	Areca nut	(d) Frond rot
385.	Oil Palm	(e) Brown Blast

	Crop	Important Pest
386.	Coconut	(a) Borer beetle
387.	Tea	(b) Rhinoceros beetle
388.	Cashew nut	(c) White and red mites
389.	Coffee	(d) Pink mite
390.	Areca nut	(e) Tea mosquito

	Crop	Important Diseases
391.	Coconut	(a) Pod Rot/Black Pod
392.	Coffee	(b) Root rot
393.	Tea	(c) Camellia dieback and canker
394.	Cocoa	(d) Leaf rust
395.	Rubber	(e) Bud rot and nutfall

	Crop	Highest Productivity
396.	Coconut	(a) Andhra Pradesh
397.	Coffee	(b) Karnataka
398.	Rubber	(c) Lakshadweep
399.	Areca nut	(d) Kerala
400.	Oil Palm	(e) Karnataka

Chapter 14

Fill in the Blanks

1. The alkaloids percent in arecanut and betel vine is ....
2. In India average productivity of plantation crops is ....
3. Largest producer country of rubber is...
4. .... is largest producer country of natural rubber.
5. India rank is ....in coffee production.
6. Mostly plantation crops cultivated in .... climate.
7. Plantation crops commercially cultivation start in..............century worldwide.
8. India share ....% of world tea exports and is also one of the...................... consumers of tea in the world.
9. India is the world's .... producer of arecanut with a share of nearly .... per cent of the total production.
10. India, Indonesia and Philippines, constitutes more than .... per cent of the global coconut production.
11. India is the............. exporter with a share of .... per cent of the global exports.
12. Arabia coffee's seeds were brought to India by ....in the 17th Century and were raised in the ....of Karnataka.
13. ....is the father of Natural Rubber.
14. In India, rubber cultivation was started in ....in..............of Kerala State, as a forest crop.
15. The first commercial plantation of rubber was started in India by European planters in .... at Always.

16. ....is largest producer of oil palm.
17. ....and .... alkaloid present in areca nut.
18. State tree of Tamil Nadu is....................
19. Indian .... community are the people who make their living from this tree using its wood, fruits, sap, stems, petioles and leaves to process a variety of food products, beverages, furniture, building materials, and handicrafts.
20. Cocoa has been known as the .... even before tea or coffee.
21. The commercial cultivation of cocoa started from ....
22. Lowering done in betel wine during ....
23. Important diseases of betel vine is......................
24. Betel vine yield about ....leaves/ha/year can be obtained.
25. Suitable temperature for cocoa growth is..................
26. Cocoa needs shade for its natural habitat, young cocoa plants grow best with .... full sunlight.
27. Flavour and aroma developed in cocoa through................process.
28. The fermentation process of cocoa is complete with in..............days.
29. Moisture contents should be below ....per cent.
30. Vitamin .... and Vitamin.............. contents are the highest in palm oil in comparison with any other types of oil.
31. Oil Palm India Limited was established in the year .... with the objective of propagating oil palm cultivation in the country and more particularly in..........
32. Palm oil content more ....acid.
33. ....is the removal of male and female flowers produced in the early stages of plantation.
34. The object of ....is to plant gain adequate stem girth, vigour and develop adequate root system.
35. The flowering starts in oil palm from ....to.........month after planting.
36. Oil Palm Development Programme (OPDP) was launched during .... under the .... with a focus on area

expansion in the States of Andhra Pradesh, Karnataka, Tamil Nadu, Orissa, Gujarat and Goa.

37. From 2004-05 onwards, the scheme is being implemented as part of the ....and provides support for Oil Palm cultivation in 12 States.

38. Fresh Fruit Bunches (FFBs) of oil palm are highly perishable and need to be processed within .... hrs of harvest.

39. Special Programme on Oil Palm Area Expansion (OPAE) was started in ....year.

40. It was suggested that palms aged 8-14 years should retain .... leaves per spiral.

41. Oil palm is ....pollinated.

42. Commercial cultivation of tea started in India from ....era when one Robert Bruce in 1823 discovered tea plants growing wild in ....valley.

43. Total area in India under is.....................thousand ha.

44. Total production of tea in India is ....millions kg.

45. Total area in Assam under is.....................thousand ha.

46. Total production of tea in Assam is ....millions kg.

47. Selection 1 (S288) variety of coffee is tetraploid hybrid from natural cross of ....

48. India produces about ....per cent of world's coffee.

49. Brazil produces about ....per cent of world's coffee.

50. Karnataka is the largest producer accounting for about .... per cent of total coffee production and ....per cent of the area under coffee in India.

51. The under and over ripe fruits can be converted into ....coffee.

52. Origin of oil palm is ....

53. ....is subfamily of oil palm.

54. Leaf production rate is between ......to.........leaves per month.

55. ....is developed by crossing between Dura x Pisifera.

56. Commercial oil palms have an economic life span of ....to........years.

57. Plateau phase of oil palm is ....to ....years.
58. Golconda is variety of...........
59. India ....per cent export coffee of total production of coffee.
60. ....coffee is highest market value.
61. ....coffee is suitable for growing at higher altitudes.
62. Harvesting time of Arabica is between................to ....
63. Harvesting time of Robusta is between................to ....
64. ....coffee is more susceptible to pests & diseases.
65. Desuckering done in coffee during ....
66. *Xylotrechus quadripes* is important insect of ....
67. ....is known as 'tree of heaven'
68. ....is known as 'Kalpavriksha'
69. Kerala, Tamil Nadu, Karnataka and Andhra Pradesh are the major coconut producing states in India accounting for more than ....per cent of area and production.
70. Swarnamangala is variety of ....
71. Seed nut should be fully ripe and having weight of above .... selected.
72. .... month old seedlings of arecanut are to be selected and transplanted in the main field.
73. .... is good time for planting of arecanut.
74. .... is highly susceptible for sun scorching.
75. Sun scorching is mostly seen in arecanut during.................. Months.
76. ....is known as 'food of god'.
77. 'Kole Roga' or 'Mahali' is important diseases of ....
78. In south India cocoa is mostly grown in ....&............garden as an intercrop.
79. Dried and fully fermented fatty seed of the cacao used in .... making.
80. Drink is prepared from cocoa is known as ....

81. Under arecanut .... of sunlight penetrates through their canopy.
82. NC-45/53 is variety of cocoa released from...............
83. ....is commercial propagation method of cocoa.
84. In cocoa .... type pruning, diseased or unnecessary branches are removed.
85. In cocoa structural pruning is done to shape the canopy to .... and ....
86. The arecanut palm start bearing from .... years.
87. For production of chali from arecanut the ....nuts are harvested from the tree.
88. Cashew was introduced in India on the coasts of Goa in 1570 AD by the Portuguese for..................
89. Cashew is considered as.......................
90. Cashew is pollinated mostly by ....
91. Coconut derives from the .... word coco, meaning "a grin".
92. The pH of coconut water is.............
93. In India per capita consumption of tea is.................
94. The main objective of withering is to .... from the leaves.
95. On an average 100 kg of fresh leaf produces .... kg of dried tea.
96. Tea leaves are partially fermented is known as ....
97. Tripura was declared the .... by the Rubber Board.
98. Root cuttings are used for .... propagation.
99. Removal of mucilage in coffee bean by .... Treatment
100. The fresh sap contains .... saccharose.

Chapter 15

Key Answers

Multiple Choice Questions

1	2	3	4	5	6	7	8	9	10
a	b	d	b	c	b	a	d	c	c
11	12	13	14	15	16	17	18	19	20
a	b	c	d	d	b	a	b	c	a
21	22	23	24	25	26	27	28	29	30
b	d	d	a	b	c	c	c	c	d
31	32	33	34	35	36	37	38	39	40
a	b	a	a	d	a	a	d	b	b
41	42	43	44	45	46	47	48	49	50
a	c	a	b	c	d	b	a	b	b
51	52	53	54	55	56	57	58	59	60
c	b	c	d	b	b	c	c	b	c
61	62	63	64	65	66	67	68	69	70
d	d	b	a	b	b	a	c	a	b
71	72	73	74	75	76	77	78	79	80
b	b	a	b	a	c	d	a	c	a
81	82	83	84	85	86	87	88	89	90
b	a	b	b	c	c	a	a	a	c
91	92	93	94	95	96	97	98	99	100
a	c	a	c	b	c	d	d	b	c

101	102	103	104	105	106	107	108	109	110
d	b	d	b	d	b	b	c	b	d
111	112	113	114	115	116	117	118	119	120
d	d	d	c	a	b	d	d	a	d
121	122	123	124	125	126	127	128	129	130
a	a	b	d	c	c	d	c	a	d
131	132	133	134	135	136	137	138	139	140
b	d	a	b	a	d	d	b	b	c
141	142	143	144	145	146	147	148	149	150
c	c	c	c	a	d	b	b	a	d
151	152	153	154	155	156	157	158	159	160
b	a	c	d	d	a	d	b	c	a
161	162	163	164	165	166	167	168	169	170
b	c	d	a	d	a	c	b	c	d
171	172	173	174	175	176	177	178	179	180
b	a	d	c	b	a	c	b	d	b
181	182	183	184	185	186	187	188	189	190
a	c	d	c	d	b	c	a	b	a
191	192	193	194	195	196	197	198	199	200
d	c	c	b	a	c	d	a	c	b
201	202	203	204	205	206	207	208	209	210
a	b	c	c	a	c	d	c	a	d
211	212	213	214	215	216	217	218	219	220
d	b	c	b	a	b	a	d	b	c
221	222	223	224	225	226	227	228	229	230
a	c	b	c	a	b	c	c	a	d
231	232	233	234	235	236	237	238	239	240
a	d	b	c	a	d	c	b	c	d
241	242	243	244	245	246	247	248	249	250
a	b	d	d	c	b	a	c	c	d

251	252	253	254	255	256	257	258	259	260
c	a	b	a	d	c	c	b	a	d
261	262	263	264	265	266	267	268	269	270
a	d	c	b	c	c	a	d	a	d
271	272	273	274	275	276	277	278	279	280
b	b	b	d	b	c	a	d	c	a
281	282	283	284	285	286	287	288	289	290
c	d	a	b	a	d	c	c	b	d
291	292	293	294	295	296	297	298	299	300
a	c	d	b	d	d	d	a	b	c

Matching

301	302	303	304	305	306	307	308	309	310
e	c	b	d	b	a	f	d	e	c
311	312	313	314	315	316	317	318	319	320
c	e	a	b	d	c	a	e	b	d
321	322	323	324	325	326	327	328	329	330
c	a	d	e	c	c	d	a	e	b
331	332	333	334	335	336	337	338	339	340
b	e	d	c	a	c	a	e	b	d
341	342	343	344	345	346	347	348	349	350
e	d	a	b	c	c	a	b	e	d
351	352	353	354	355	356	357	358	359	360
e	c	b	a	d	e	d	a	c	b
361	362	363	364	365	366	367	368	369	370
b	e	d	c	a			a	b	c
371	372	373	374	375	376	377	378	379	380
b	d	e	c	a	c	e	b	a	d
381	382	383	384	385	386	387	388	389	390
c	d	e	a	b	b	d	e	a	c
391	392	393	394	395	396	397	398	399	400
e	d	c	a	b	b	c	d	e	a

Fill in the Blanks

1. Acrolein
2. 4.7 MT/ha
3. Brazil
4. India
5. 6th
6. Tropical
7. 18th
8. 1st, largest
9. Leading, 50%
10. 75%
11. Leading, 65%
12. Baba Budan, Baba Budan Hills
13. Si. Henry Wickham
14. 1878, Nilambur
15. 1902
16. Malaysia
17. Arecaidine, arecoline
18. Palmyra Palm
19. Nadar
20. Beverage crop
21. 1960's
22. March – April
23. Phytophthora Wilt
24. 75 to 100 lakh
25. 25°C
26. 50%
27. Fermentation
28. 3-5
29. 6-8
30. A, E
31. 1977, Kerala
32. Palmitic acid
33. Ablation
34. Ablation
35. 14th , 18th
36. 1991-92, Technology Mission on Oilseeds and Pulses" (TMOP)
37. Integrated Scheme of Oilseeds, Pulses, Oil Palm & Maize (ISOPOM)
38. 24
39. 2011-12
40. 5-6
41. Wind
42. British, Upper Brahmaputra
43. 563.98
44. 1208.78
45. 304.40
46. 629.05
47. C. liberica x C. Arabica
48. 2.5
49. 25
50. 70, 60
51. Cherry
52. West Africa
53. Cocoidae
54. 1, 3
55. Tenera
56. 20, 30
57. 10, 15
58. Tea
59. 65-70
60. Arabica
61. Arabica
62. November to January
63. December to February
64. Arabica
65. June-July
66. Coffee
67. Coconut
68. Coconut
69. 90
70. Arecanut
71. 35 g
72. 18
73. September-October

74. Arecanut
75. October – January
76. Cocoa
77. Areca nut
78. Arecanut, Coconut
79. Chocolate
80. Hot chocolate
81. 30-50%
82. Central Plantation Crops Research Institute (CPCRI), Kasaragod, Kerala
83. Soft wood grafting.
84. Sanitary
85. Desired size, Architecture
86. 4-5
87. Ripened
88. Soil conservation
89. "a waste land crop"
90. Insects
91. Spanish
92. 4.5
93. 0.3 kg
94. Removing of moisture
95. 22.5
96. Oolong Tea
97. "Second Rubber Capital of India"
98. Robusta coffee
99. Alkali
100. 12%

NIPA Publications on Horticulture

S.No.	Title	Author	ISBN	Year
1	A Colour Handbook on Practical Plant Pathology	Yadav, Vijay	9789385516177	2015
2	A Colour Handbook: Landscape Gardening	Singh, Alka	9789383305889	2015
3	A Handbook of Minerals,Crystals,Rocks and Ores	Alexander, P.O.	9788190723787	2009
4	A Handbook of Soil-Plant-Water-Fertilizer and Manure Analysis	Durai, M.V.	9789381450185	2014
5	A Handbook on Irrigation and Drainage	Panigrahi, Balram	9789381450888	2013
6	A Text Book of Seed Science and Technology	Padmavathi, S.	9789381450437	2012
7	Abiotic Stress Tolerance in Crop Plants: Breeding and Biotechnology	Roy, Bidhan	9788189422943	2009
8	Acid Soils: Their Chemistry and Management	Sarkar,A.K.	9789381450383	2013
9	Advances and Challenges in Agricultural Extension and Rural Development	Rathakrishnan,T.	9789380235035	2009
10	Advances in Preservation and Processing Technologies of Fruits and Vegetables	Rajarathnam, S.	9789380235523	2011
11	Advances in Protected Cultivation	Singh,Brahma, Balraj Singh	9789383305179	2014
12	Advances in Soil Borne Plant Diseases	Naik, Manjunath	9788189422813	2008
13	Agricultural Biotechnology: Indian Print	Persley, G.J.: et al	9788189422097	2006
14	Agricultural Extension: Worldwide Innovations	Saravanan, R.	9788189422967	2008
15	Agricultural Marketing: Perspectives and Potentials	Bhat, Anil	9789385516153	2016
16	Agricultural Plant Biochemistry	Nagaraj, G.	9789383305551	2015
17	Agricultural Statistics: A Guide for Competitive Examinations	Kushwaha, K.S.	9789381450314	2012
18	Agriculture Bioinformatics	Keshavachandran, R.	9789383305421	2015
19	Agri-Food Crops: Processing,Value Addition, Packaging and Storage	R.Sasi Kumar	9789381450406	2012
20	Agri-Horticultural Biodiverstiy of Temperate and Cold Arid Regions	Zeerak,NazirAhmad	9789381450086	2012
21	Agrobiodiversity and Sustainable Rural Development	S.K. Soam	9789383305896	2015
22	Agro-enterprises for Rural Development and Livelihood Security	Sharma J.P. & S.K.Dubey	9789380235851	2011
23	Agroforestry Systems for Resource Conservation and Livelihood Security in Lower Himalays	Panwar,P & Dadhwal	9789381450215	2012
24	Agroforestry: Principles and Practices	Patra, Alok Kumar	9789381450765	2013
25	Agroforestry: Systems and Practices	Puri, Sunil & Pankaj Panwar	9788189422622	2007
26	Agroforestry: Systems and Prospects	C.B.Pandey & O.P.Chaturvedi	9789381450970	2014
27	Agro-Informatics	Vanitha, G & M.Kalpana	9789380235714	2011
28	Agrotechnology Manual: Including Nursery Management and Practices	Etomes, Marcel N	9789383305018	2014
29	An Introduction to Intellectual Property Rights	Pathak, Manju	9789383305124	2014
30	An Introduction to Nanotechnology	Rathinasamy, A.	9789381450413	2012
31	Ancestral Knowledge in Agri-Allied Science	Saha, Ratan Kumar	9789383305216	2014
32	Applied Computational Biology and Statistics in Biotechnology and Bioinformatics (Set of 2 Vols.)	Roy, A.K.	9789380235929	2012
33	Applied Statistical Techniques	Imam,Ekwal	9789383305537	2015

S.No.	Title	Author	ISBN	Year
34	Applied Statistics for Agricultural Sciences	D.Venkatesan	9789383305285	2014
35	Approaches for Incorporating Drought and Salinity Resistance in Crop Plants	Chopra,V.L. & R.S.Paroda	9789383305742	2015
36	Aromatic Plants: Vol.01. Horticulture Science Series	Skaria, B.P. et.al.	9788189422455	2007
37	Auddaniki ke Adharbhoot Sidhant Tatha Nashijeev Prabandhan	Tiwari, A.K.	9789380235998	2012
38	Bananas and Plantains: Postharvest Management, Storage,Ripening and Processing	Narayana, C.K.	9789383305452	2015
39	Basic Concepts in Statistics	Kushwaha, K.S.	9788189422400	2009
40	Basics of Horticulture: 2nd Revised and Expanded ed. (As per Revised ICAR Syllabus)	Peter, K.V.	9789383305735	2015
41	Basics of Mutation Breeding	Thirugnanakumar,S.	9789383305193	2014
42	Basics of Research Methodology	Imam,Ekwal	9789383305438	2015
43	Biochemical Aspects of Plant Physiology	Bhattacharya, A. & Vijaya Luxmi	9789383305902	2016
44	Biochemistry,Molecular Biology and Biotechnology: Instant Notes	Gajera, H.P.,	9789383305520	2015
45	Bioinformatics in Agriculture: Tools and Applications	Balakrishnan,M.:Ed.	9789381450925	2014
46	Biostatistics: Basic Concepts and Methodology	Kushwaha, K.S.	9789383305469	2014
47	Biotechnology in Horticulture: Methods and Applications	Peter, K.V.	9789381450918	2013
48	Biotechnology in India: Initiatives and Accomplishments	Niladri Bag ed.	9789385516252	2016
49	Biotechnology: Practical Manual Series Vol 04	Thara, K.M.	9788190851237	2009
50	Breeding and Biotechnology of Flowers: Set of 2 Vols.	Singh, A.K.	9789383305612	2014
51	Breeding and Biotechnology of Flowers: Vol.1: Commercial Flowers	Singh, A.K.	9789383305353	2014
52	Breeding and Biotechnology of Flowers: Vol.2: Garden Flowers	Singh, A.K.	9789383305407	2014
53	Breeding and Protection of Vegetables	Rana, M.K.	9789380235493	2011
54	Breeding of Horticultural Crops: Principles and Practices: 2nd Revised & Expanded ed. (As per Revised ICAR Syllabus)	Kumar,N.	9789383305773	2015
55	Breeding,Biotechnology and Seed Production of Field Crops	Roy, Bidhan	9789381450680	2013
56	Cashew Production & Processing Technology: Recent Advances	Gajbhiye, R.C.	9789383305827	2015
57	Climate Change and Agricultural Food Production	Kibria, Golam et.al.	9789381450512	2013
58	Climate Change and Food Security	Datta, M.& N.P.Singh	9788189422387	2008
59	Climate Change and Natural Resources Management	Lenka,S.&N.K.Lenka	9789381450673	2013
60	Climate Mitigation and Carbon Finance: Global Initiavities & Challenges	Sahoo, A.K.	9789381450024	2012
61	Climate Resilient Crops for the Future	Peter,K.V.	9789383305599	2015
62	Climatic Variability: Impacts on Agriculture and Allied Sectors	Datta, M.:Ed.	9789381450949	2014
63	Commercial Crops Technology: Vol.08. Horticulture Science Series	Kurian, Alice & K.V. Peter	9788189422523	2007
64	Commercial Horticulture	Patel & Chawla	9789385516238	2016
65	Computers in Agriculture: Fundamentals and Applications	Sharma Manish & Anil Bhatt	9789385516160	2015
66	Crop Diseases Management: Principles and Practices	Narayanasamy, P.	9789380235677	2011
67	Crop Diseases: Identification,Treatment and Management	Darwin, Henry	9789380235462	2011
68	Developments in Physiology,Biochemistry and Molecular Biology of Plants Vol 01	Bose, Bandana & A.Hemantaranjan: ed.	9788189422028	2005
69	Developments in Physiology,Biochemistry and Molecular Biology of Plants Vol 02	Bose, Bandana & A.Hemantaranjan: ed.	9788189422929	2008

S.No.	Title	Author	ISBN	Year
70	Dimensions of Extension Education	Mohapatra, B.P.	9789381450987	2014
71	Diseases of Horticultural Crops Identification and Management: With Colour Illustrations	Kumar, Sanjeev	9789383305643	2015
72	Diseases of Vegetable Crops and Their Integrated Managment: A Colour Handbook	Mishra, R.K.	9789381450499	2013
73	Drip and Sprinkler Irrigation	Biswas,R. Kumar	9789383305766	2015
74	Droughts in Agricultural Production: Monitoring and Management	G.G.S.N.Rao	9789385516009	2015
75	Drug Discovery and Development: Traditional Medicine and Ethnopharmacology	Patwardhan,Bhushan	9788189422295	2007
76	Ecologically Based IPM (Set of 2 Vols.)	Abrol& Shankar	9789380235950	2012
77	Economics,Marketing and Sales of Agricultural Products	Thakur S.N.	9789381450659	2013
78	Efficiency Indices for Agriculture Management Research	Devasenapathy, P. & B.Gangwar	9788189422882	2008
79	Elements of Entomology	H.L.Devasahayam	9789381450635	2014
80	Emerging Technologies of the 21st Century	Roy,A.K.	9789383305339	2015
81	Encyclopedia of Agricultural Meteorology	Dhaliwal, L.K. & S.S.Hundal	9789380235547	2011
82	Entomology: Novel Approaches	Jain,P.C. & M.C.Bhargava	9788189422325	2007
83	Essential Oils and their Applications	Das, Kuntal	9789381450741	2013
84	Ethnomedicinal Plants Resource of Orissa: Vol.01	Sahoo, A.K.	9789380235752	2011
85	Evaluation and Impact Assessment of Technologies and Developmental Activities in Agriculture,Fisheries and Allied Fields	Roy, A.K.	9789380235400	2011
86	Experimental Biotechnology: Practical Manual Series 06	Dutta, Sunita & Abhijit Dutta	9789380235721	2011
87	Experimental Phytochemical Techniques	Raaman, N.	9789380235943	2012
88	Extension Management in the Information Age Initiatives and Impacts	Philip, H. & T.Rathakrishnan	9789381450543	2013
89	Extension of Technologies: From Labs to Farms	Anandaraja, N.	9788189422837	2008
90	Family Farming and Rural Economic Development	M.L.Choudhary & Aditya	9789383305858	2015
91	Flower Crops: Cultivation and Management	Singh, A.K.	9788189422356	2006
92	Flowering Trees: Vol.12. Horticulture Science Series	Valsalakumari, P.K.	9788189422509	2008
93	Flowers for Trade: Vol.10. Horticulture Science Series	Sheela, V.L.	9788189422516	2008
94	Food and Nutritonal Security by Sustainable Agriculture: Methods to Attain and Sustain	Mishra, Kumar Bijesh	9789383305049	2014
95	Food Engineering and Technology	Sharma, H.K. & Ashutosh Upadhyay	9789383305483	2015
96	Food Process Engineering and Technology	Pare, A.&Mandhyan	9789380235431	2011
97	Food Processing Waste Management: Treatment and Utilization Technology	Joshi V.K. & Satish Sharma	9789380235592	2011
98	Food Science	Bawa, A.S. & O.P.Chauhan	9789381450147	2013
99	Food Science and Technology: Glossary of Preeminence	Dev Raj	9789380235806	2011
100	Fruit Breeding	Dinesh, M.R.	9789383305513	2015
101	Fruit Crops: Vol.03. Horticulture Science Series	Radha,T. & Mathew	9788189422462	2007
102	Functional Foods	Mishra, H.N.	9789383305988	2016
103	Functional Foods and Nutraceuticals	C.S. Riar, et al.	9789383305964	2015
104	Fundamentals of Garden Designing: A Colour Encyclopedia	Roy, Rup Kumar	9789381450307	2013

S.No.	Title	Author	ISBN	Year
105	Fundamentals of Ornamental Horticulture and Landscape Gardening	Tiwari, A.K.	9789381450079	2012
106	Fundamentals of Vegetable Production	Rana, M.K.	9789380235707	2011
107	Genetic Resources and Seed Enterprises: Management and Policies: In 2 Parts	Ram, Hari Har & R.Yadava	9788189422653	2007
108	Genomics and Genetic Engineering	Satya, Pratik	9788189422776	2007
109	Geographic Information System	Gurugnanam, B.	9788190851282	2009
110	Geoinformatics Applications in Agriculture	Singh,Anil Kumar	9788189422233	2008
111	Geospatial Technologies for Natural Resources Management	Soam,S.K. & P.D. Sreekant	9789381450802	2013
112	GIS: Fundamentals,Applications and Implementations	Elangovan, K.	9788189422165	2006
113	Heterosis Breeding in Vegetable Crops	Rai, Nagendra	9788189422035	2006
114	Horticulture Science Series: Vol.01-12: Set of 12 Vols.	Peter, K.V.: Series Ed.	9788189422479	2008
115	ICTs for Agricultural Extension: Global Experiments, Innovations and Experiences	Saravanan, R. ed.	9789380235240	2010
116	ICTs for Transfer of Technologies: Tools & Techniques	Verma, S.R.	9788193014479	2015
117	Identification and Management of Horticultural Pests	Ranjith, A.M.	9789381450567	2013
118	Illustrated Dictionary of Entomology	Paras Nath	9788189422561	2007
119	Illustrated Plant Pathology: Basic Concepts	Darwin, Henry	9789380235080	2009
120	Improving Productivity of Drylands by Sustainable Resource Utilization and Management	Dayal, Devi	9789385516191	2016
121	Indigenous Medicinal Plants and their Practical Utility	Lakshman, H.C.	9789381450116	2012
122	Information Access in Digital Libraries	Stanley Madan Kumar	9789383305384	2014
123	Information and Communication Technology for Agriculture and Rural Development	Saravanan, R.	9789380235882	2011
124	Information and Knowledge Management: Tools, Techniques and Practices	Roy, A.K.	9789381450628	2013
125	Innovative Horticulture	Arunkumar, K.	9788189422738	2008
126	Integrated Farming System Practices: Challenges and Opportunities	Nanda, S.S.	9789385516207	2016
127	Integrated Pest Management in the Tropics (In 2 Parts)	Abrol,Y.P.& Shankar	9789385516115	2016
128	Intellectual Property Rights Demystified	Ramkumar, Mu.	9788189422875	2008
129	IPR: Drafting,Interpretation of Patent Specifications and Claims	Rathore,N.S.: Ed.	9789381450819	2013
130	Irrigation and Agricultural Drainage Engineering	Biswas, R. Kumar	9789383305247	2015
131	Irrigation Systems Engineering	Panigrahi, Balram	9789380235387	2011
132	Laboratory Manual of Biochemistry: Methods and Techniques	Sengar, R.S.	9789383305025	2014
133	Laboratory Manual of Microbiology: Practical Manual Series: 05	Roy, A.K.	9789380235189	2010
134	Library Services in the Knowledge Web	Veeranjaneyulu & R.Mahapatra	9789381450192	2012
135	Management of Horticultural Crops: Vol.11 Horticulture Science Series: (Part-I & II Combined In 1 Binding)	Pradeepkumar, T.	9788189422493	2008
136	Mechanization of Cultivated Crops	Singh,Surendra	9789383305759	2015
137	Medicinal Plants: Holistic Approaches	Zawar, Nudrat Sayed	9789380235585	2011
138	Medicinal Plants: Vol.02. Horticulture Science Series	Kurian, A. & A.Sankar	9788189422424	2007
139	Merging Plant Breeding with Crop Biotechnology	Yasin, J.K.:Ed.	9789381450598	2012
140	Methods and Techniques in Plant Physiology	Bhattacharya, A. & Vijaya Luxmi	9789383305506	2015
141	Microbial Biotechnology for Sustainable Agriculture, Horticulture and Forestry	Bagyaraj, D.J.	9789380235820	2011
142	Microbial Diversity and Functions	Bagyaraj, & Tilak,	9789381450109	2012

S.No.	Title	Author	ISBN	Year
143	Microbial Diversity and Its Applications	Barbudde, S.B. R.Ramesh & Singh	9789381450666	2013
144	Mobile Phones for Agricultural Extension	Saravanan,R.	9789383305230	2014
145	Modern Biotechnology and Its Applications (Set of 2Vols.)	Behera, K.K.	9789381450833	2013
146	Modern Methods in Plant Physiology	Srivastava, G. Chand	9789380235011	2010
147	Modern Technologies for Sustainable Agriculture	Kumar, Sunil	9789381450611	2013
148	Modern Technology in Vegetable Production	Hazra, Pranab	9789380235325	2011
149	Molecular Markers and Plant Biotechnology	Tomar, R.S.	9789380235257	2010
150	Nanotechnology in Agriculture	Subramanian, K.S.	9789383305209	2015
151	Nanotechnology in Soil Science and Plant Nutrition	Adhikari, Tapan	9789381450789	2013
152	Nematology: Fundamentals and Applications	Jonathan, E.I.	9789380235141	2010
153	Numericals and Short Questions in Farm Machinery, Power and Energy in Agriculture	Yadav, Rajvir	9788190723718	2009
154	Nutritional Disorders in Fruit Crops: Diagnosis and Management	Prakash, M.	9789381450956	2013
155	Objective Horticulture	Verma, Anil	9789381450239	2012
156	Oilseeds: Properties,Products,Processing and Procedures	Nagaraj, G.	9788190723756	2009
157	Orchid: Cultivation and Management	Tiwari, A.K.	9789383305001	2014
158	Organic Farming	Singh, A.K.	9789385516139	2015
159	Organic Farming Scope and Uses of Biofertilizers	Panwar, JDS & Amit Jain	9789385516184	2016
160	Organic Spices	Parthasarathy, V.A	9788189422844	2008
161	Ornamental Horticulture	Sindhu, & Peter, K.V.	9789383305865	2016
162	Ornamental Plants	Sabina, George	9788190851268	2009
163	Pest Management and Residual Analysis in Horticultural Crops	Gulati Rachna & Beena Kumari	9789381450710	2013
164	Pesticides: Methods of Their Residues Estimation	Kumari, Beena & T.S.Kathpal	9789380235394	2010
165	Pests of Vegetables: Bionomics and Management	Laskar, Nripendra	9789385516016	2015
166	Phal Evam Sabji Parirakshan Digdarshika	Kumari, Arunima	9789380235561	2011
167	Physio-Biochemistry and Biotechnology of Vegetable Crops	Rana, M.K.	9789380235318	2011
168	Physio-Biochemistry and Biotechnology of Vegetable Crops	Rana, M.K.	9789380235318	2011
169	Physiological Disorders of Fruit Crops	Savreet, Sandhu & Bikramjit Singh Gill	9789381450581	2013
170	Phytochemical Techniques	Raaman, N.	9788189422301	2006
171	Plant Biochemistry: Techniques and Procedures	Nagaraj, G.	9789383305940	2015
172	Plant Diseases: Identification and Management (With Illustrations)	Rai, J.P.	9789383305315	2014
173	Plant Nutrient Disorders: Diagnosis and Management	Sarkar, & Mahapatra	9789385516023	2015
174	Plant Pathogens and Principles of Plant Pathology	Singh, Sanjeev	9789385516078	2016
175	Plant Secondary Metabolities	Shukla & Bhatnagar	9788190851220	2009
176	Plant Toxonomy and Biosystematics: Classical and Modern Methods	Rana,T.S.	9789383305414	2014
177	Postharvest Management and Processing of Fruits and Vegetables: Instant Notes	Sharma, Satish	9789380235202	2010
178	Postharvest Techniques and Management for Dry Flowers	V. Ponnuswami & Aruna,P.	9789380235868	2011
179	Postharvest Technologies for Commercial Floriculture	Verma, Anil	9789381450048	2012
180	Postharvest Technology and Engineering: An Illustrated Guide	Dev Raj	9789381450451	2012

S.No.	Title	Author	ISBN	Year
181	Postharvest Technology and Processing of Horticultural Crops	Pandit, P.S.	9789383305278	2014
182	Postharvest Technology of Horticultural Crops: Practical Manual Series Vol 02	Sharma, Satish & M.C.Nautiyal	9788190851206	2009
183	Postharvest Technology of Horticultural Crops: Vol.07. Horticulture Science Series	Sudheer, K.P. & V.Indira	9788189422431	2007
184	Practical Manual of Entomology (Insects and Non-Insects Pests)	D.H.Lewin	9789380235905	2011
185	Practical Manual of Horticulture Crops: Set of 2 Vols.	Verma, Anil Kumar	9789383305711	2015
186	Practical Manual of Horticulture Crops: Vol.01: Production Technologies	Verma,Anil Kumar	9789383305704	2015
187	Practical Manual of Horticulture Crops: Vol.02: Processing and Postharvest Technologies	Vaidya,Devina	9789383305698	2015
188	Practical Plant Biotechnology and Genetics	Rani, Archana	9789383305995	2015
189	Precision Farming in Horticulture	Singh, Jitendar, S.K.Jain,L.K.Dashora	9789381450475	2013
190	Production Technology of Spices, Aromatic, Medicianl and Plantation Crops	Barche, Swati	9789385516061	2016
191	Propagation of Horticultural Crops: Vol.06. Horticulture Science Series	Rajan, S.& B.L.Markose	9788189422486	2007
192	Propagation of Horticultural Plants: Arid and Semi-Arid Regions	Singh, R.S. & R.Bhargava	9789383305254	2014
193	Protected Cultivation of Horticultural Crops	Singh, D.K & K.V.Peter	9789383305155	2014
194	Quality Control for Value Addition in Food Processing	Dev Raj, R.Sharma & Joshi	9789380235578	2011
195	Quantitative Genetics and Crop Breeding	Thirugnanakumar, S.	9789380235981	2012
196	Question Bank in Extension Education	Mohammad,Asif	9788193014493	2015
197	Question Bank in Postharvest Technology	Guleria, SPS & Anil Kumar Verma	9789380235042	2010
198	R Statistics	Samuel, Duleep	9789385516146	2015
199	Recent Advances in Biopesticides	Johri, Jayendra: eds.	9789380235219	2010
200	Remember Your Humanity: Pathway to Sustainable Food Security	Swaminathan, M.S.	9789381450178	2012
201	Rural Livelihood and Food Security	Wani, M.H.	9789380235936	2012
202	Seed Production of Field Crops	Mondal, S.S.	9788190723763	2009
203	Seed Science and Technology	Vanangamudi,K.	9789383305117	2014
204	Soil Conservation: Fully Revised and Updated 3rd ed.	Norman Hudson	9789383305971	2015
205	Soil Microbiology and Biochemistry	Hassan, G.Dar	9789380235134	2010
206	Soil Sampling and Methods of Analysis	Pal, Sushant	9789381450574	2013
207	Soil Science: An Elementary Textbook	Puri, A.N.	9789383305063	2015
208	Soil Testing and Analysis: Plant, Water and Pesticide Residues	Brajendra, Patiram	9788189422707	2007
209	Spices, Plantation Crops, Medicinal and Aromatic Plants: A Handbook	Tyagi, S.K.	9788193014486	2015
210	Spices: Vol.05. Horticulture Science Series	Nybe, E.V. and Mini Raj	9788189422448	2007
211	Statistical Designs and Analysis for Agricultural Field Experiments	Katyal, & D.M.Hegde	9789381450840	2013
212	Statistical Methods for Agricultural Field Experiments	Katyal, Vijay & B.Gangwar	9789380235424	2011
213	Sustainable Agriculture: A Vision for Future	Desai, B.K. and B.T.Pujari	9788189422639	2007
214	Sustainable Environmental Science	Sahu, D.D.	9789381450208	2012
215	Systematics of Fruit Crops	Sharma, Girish	9789380235066	2009

S.No.	Title	Author	ISBN	Year
216	Technologies for Sustainable Green Environment	Davamani, V.	9789381450420	2012
217	Temperate Horticulture: Current Scenario	Kishore, D.K.: et.al.	9788189422363	2006
218	The Basics of Human Civilization: Food, Agriculture and Humanity Vol.01 Present Scenario	Prem Nath	9789381450734	2013
219	The Basics of Human Civilization: Food, Agriculture and Humanity Vol.02 Food	Prem Nath	9789383305377	2014
220	The Chemistry of Soil Processes	D.J. Greenland	9789383305926	2015
221	The Pomegranate	Hiwale, S.S.	9789380235158	2009
222	The Qur-anic Plants and Animals	Zeerak, Nazir Ahmad	9789383305780	2015
223	The Science of Horticulture Vol 01	Peter, K.V.: eds.	9789380235479	2011
224	The Science of Horticulture Vol 02	Peter, K.V.: eds.	9789380235486	2011
225	The Theory of Sample Surveys and Statistical Decisions	Kushwaha, K.S. & Rajesh Kumar	9788189422899	2009
226	Traditional Agricultural Practices: Applications and Technical Implementations	Rathakrishnan,T.	9789380235028	2009
227	Tuber and Root Crops: Vol.09. Horticulture Science Series	Palaniswami,M.S.& K.V.Peter	9788189422530	2008
228	Turning Plants Into Medicines: Novel Approaches	T.Parimelazhagan	9789381450468	2013
229	Underutilized and Underexploited Horticultural Crops: Vol 01	Peter, K.V.: ed.	9788189422608	2007
230	Underutilized and Underexploited Horticultural Crops: Vol 02	Peter, K.V.: ed.	9788189422691	2007
231	Underutilized and Underexploited Horticultural Crops: Vol 03	Peter, K.V.: ed.	9788189422851	2008
232	Underutilized and Underexploited Horticultural Crops: Vol 04	Peter, K.V.: ed.	9788189422905	2008
233	Underutilized and Underexploited Horticultural Crops: Vol 05	Peter, K.V.: ed.	9789380235288	2010
234	Value Addition in Flowers and Orchids	De, L.C.	9789381450123	2011
235	Vegetable Crops: Genetics Resources and Improvements	D.K.Singh & H.Choudhary	9789380235509	2012
236	Vegetable Crops: Vol.04. Horticulture Science Series	Gopalakrishnan, T.R.	9788189422417	2007
237	Vegetable Seed Processing	Mathad, R.C.	9789385516030	2015
238	Weed Science	Das, P.C.	9789383305261	2015

FSC
www.fsc.org
MIX
Papier | Fördert gute Waldnutzung
FSC® C083411

Zeitfracht Medien GmbH
Ferdinand-Jühlke-Straße 7
99095 Erfurt, Deutschland
produktsicherheit@kolibri360.de